HIGH GRADIENT ACCELERATING STRUCTURE

Proceedings of the Symposium on the
Occasion of 70th Birthday of Juwen Wang

HIGH GRADIENT ACCELERATING STRUCTURE

Tsinghua University, China 19 May 2013

W Gai

Argonne National Laboratory, USA

NEW JERSEY • LONDON • SINGAPORE • BEIJING • SHANGHAI • HONG KONG • TAIPEI • CHENNAI

Published by

World Scientific Publishing Co. Pte. Ltd.
5 Toh Tuck Link, Singapore 596224
USA office: 27 Warren Street, Suite 401-402, Hackensack, NJ 07601
UK office: 57 Shelton Street, Covent Garden, London WC2H 9HE

British Library Cataloguing-in-Publication Data
A catalogue record for this book is available from the British Library.

HIGH GRADIENT ACCELERATING STRUCTURE
Proceedings of the Symposium on the Occasion of 70th Birthday of Juwen Wang

ISBN 978-981-4602-09-9

In-house Editor: Christopher Teo

Printed in Singapore

Preface

This special symposium was held at Tsinghua University to report on recent progress in the development of the high gradient RF accelerator and to celebrate the 70th birthday of Dr. Juwen Wang. In the past several decades, Dr. Wang has made many monumental contributions to the high gradient RF accelerator studies in theory, experiments, and development of technology. We have all benefited from his wisdom and guidance, as illustrated in the papers contained within the following proceedings.

More than 50 people attended the symposium. We were especially honored with the participation and wonderful presentations of Drs. Loew and Miller. The organizing committee is very grateful for the hospitality and financial contributions from the department of Engineering Physics at Tsinghua University. We would also like to acknowledge the fine work of the conference secretary, Ms. Xue Fan, as well as that of other local services. I would also like to thank Mr. Jiahang Shao for helping me edit these proceedings.

On behalf of the organizing committee, I would like to thank all the participants for their contributions. This was a wonderful event and we wish Dr. Wang a happy birthday and best wishes in the future.

Wei Gai
Conference Proceedings Editor
Chuanxiang Tang
Symposium Chair

Contents

Juwen Wang and High-Gradient LINACS a Celebration at Tsinghua University

Gregory Loew
SLAC National Accelerator Laboratory
Menlo Park, CA 94025, U. S. A.

1. Introduction

My report for this celebration comes in three parts:

1) Juwen Wang's early history,
2) His Ph.D. thesis,
3) His work since 1989.

2. Juwen Wang's Early History

Juwen Wang was born in Tai Kang County, He Nan Province, China on July 28th, 1943. His parents were Guozhang Wang and Yunqin Ji. He had two sisters, Juyan and Jufang, and one brother, Juwu who, after their parents' deaths, was instrumental in bringing up Juwen and supporting him. After his high school education, Juwen entered Tsinghua University for the first time in 1961. In 1968, during the Cultural Revolution, he left Tsinghua and went to work at a railway factory in Xi'an where he built electronics equipment for railways until 1978. In Xi'an, Juwen met Xiuqin Yuan and they were married on March 8th, 1973, forty years ago.

In 1978, after the Cultural Revolution, Juwen returned to Tsinghua University for a graduate degree in physics. The next year, he decided that he wanted to specialize in the field of particle accelerators. To realize his dream, he thought he would try to come to the United States of America to get a Ph.D. in this field. Somewhat by coincidence but also because of his persistence, Juwen and I met in September 1979 at an Accelerator School in Hefei to which I had been invited to give lectures.

Fig. 1. Xiuqin Yuan and Juwen Wang in 1973.

Fig. 2. Hefei was an amazing city where manual labor prevailed.

Juwen was an excellent student and he was held in high regard by his professors in China, in particular Professor Liu Naichuan. As a result, he was able to obtain a scholarship to study in the United States in 1980, and to come to work with me at SLAC. This event started a wonderful friendship and technical relationship between us.

During the first two years, Juwen did some very interesting work to measure the energy lost to longitudinal modes by single electron bunches traversing various periodic structures for linear accelerators, such as disk-loaded structures and alternating-spoke structures. This gave him some very valuable theoretical and experimental experience. As a result, he was admitted to the Ph.D. program in the Applied Physics Department at Stanford, which gave him the chance to take graduate level courses at the university and pursue a Ph.D. thesis with me.

Fig. 3. Student colleagues of Juwen Wang in Hefei, ready to learn.

3. Juwen Wang's Ph.D. Thesis

Juwen Wang's thesis became a very broad study of linear electron accelerators, covering the derivation of transient beam loading calculations in constant-gradient structures, wake field calculations and measurements, methods to shape electron bunches to obtain flat energy spectra, and detailed high-gradient breakdown studies in room temperature structures [1].

Because of the magnitude of this work, this report concentrates on the RF breakdown studies. Short of the old Kilpatrick breakdown criterion based on the effect of ion emission from cavity walls, the prevailing hypothesis at the time to explain RF breakdown was that it was caused by large electric surface fields resulting in excessive electron field emission, metal melting and damage. However, none of the details were well understood, and many experiments were needed.

In summary, what Juwen's thesis established was that:

1) Breakdown begins at relatively low field levels during RF processing and outgassing of a structure.
2) With care, the gradient can be gradually increased up to a much higher level, without damage to the structure's internal surface.
3) Eventually, a maximum gradient is reached, sparking appears on iris edges, and damage is inevitable.
4) At all field levels, electron field emission shows up in the form of dark current.
5) When breakdown occurs, the emitted current momentarily jumps up by a factor of about 40, and outgassing simultaneously increases abruptly. Then, the emitted current drops back to its steady-state value for that gradient.

Nobody knew at the time if there was an RF frequency dependence for breakdown. For this reason, experiments were carried out with several structures at different RF frequencies, as shown in Fig. 4 below. Figure 5 shows the experimental set-up, the probes and the analyzing magnet to measure the magnitude and energy of the emitted electron currents.

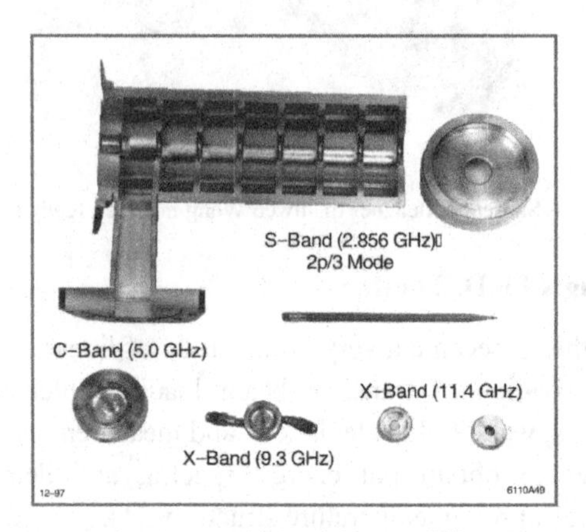

Fig. 4. Structures used to measure RF breakdown at different frequencies.

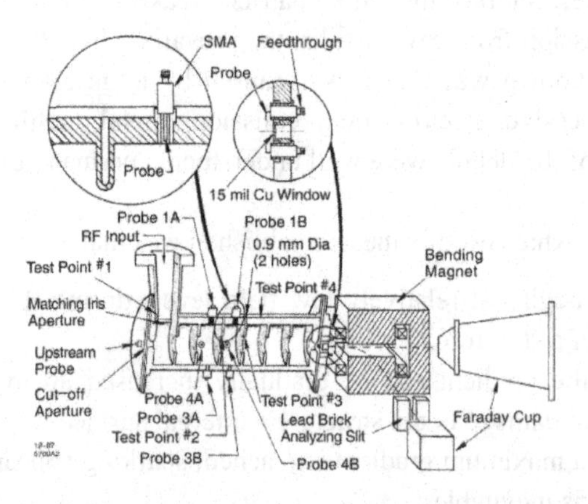

Fig. 5. Experimental set-up to measure emitted currents.

Figure 6 shows a typical set-up used to observe reflected RF signals, dark current beam spots and gas releases due to breakdown. Figure 7 shows typical gas releases during (a) steady-state operation, and (b) immediately after breakdown, where the CO^+ and CO_2^+ lines are greatly enhanced.

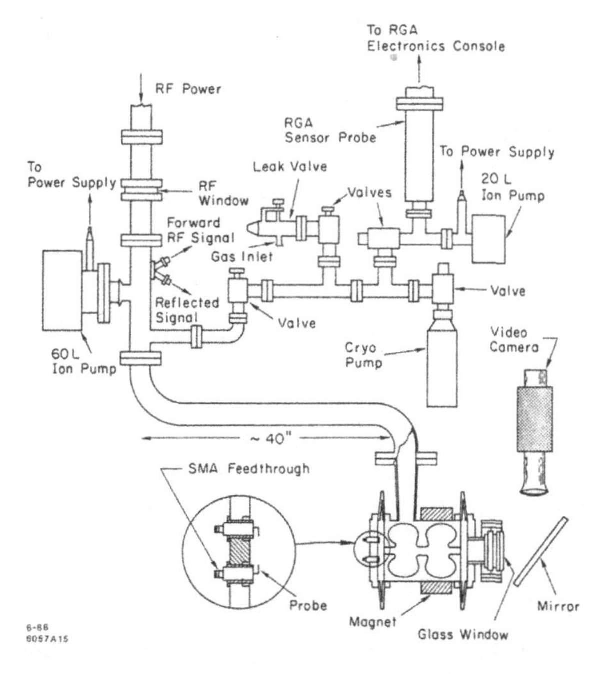

Fig. 6. Set-up to observe reflected RF signals, dark current beam spots and gas releases.

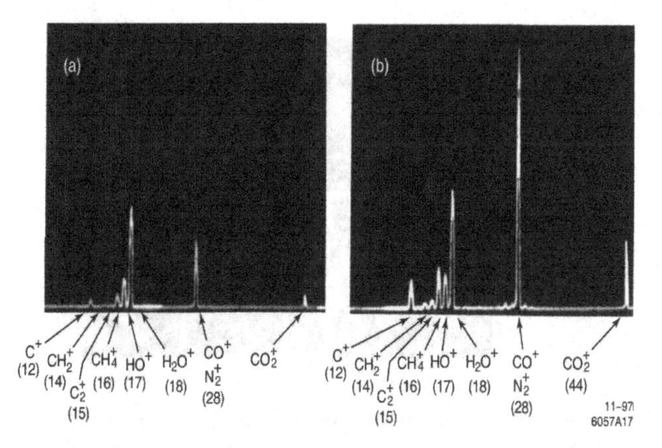

Fig. 7. Gas releases (a) in steady-state, (b) immediately after breakdown.

The surface field limits as a function of RF frequency that are shown in Fig. 8 were measured for the structures shown in Fig. 4. Note that this frequency dependence was obtained for standing-wave fields and that the filling times and RF pulse lengths available from the microwave sources were not the same for each structure. Hence, the similarity with the old Kilpatrick square-root of frequency breakdown dependence was probably purely coincidental. Also note that these field limits were attained after considerable RF processing and surface damage as shown later in Fig. 9.

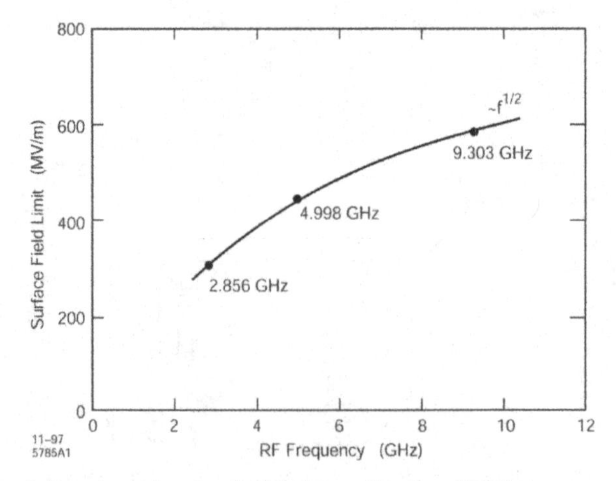

Fig. 8. Experimental surface field limits as a function of RF frequency.

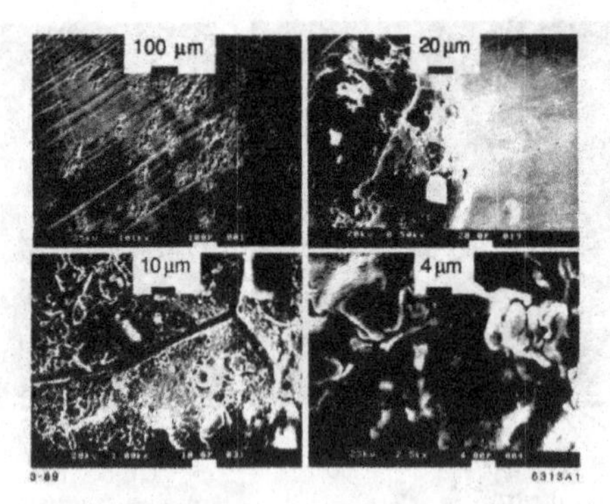

Fig. 9. S-band copper cavity wall damage after breakdown tests seen with a scanning electron microscope.

The next step in the investigation of the mechanism of RF breakdown, assuming that it was due to excessive electric surface fields, was to estimate the dependence of the field emitted current I_{FE} as a function of electric field E, surface field enhancement factor β (due to surface imperfections or impurities), effective emitting area A_e ϕ and work function ϕ. Fowler and Nordheim [2] had originally developed their theory in 1928 for dc fields and it was necessary to convert it to ac fields. Juwen did this with considerable effort and obtained the following expressions:

$$\bar{I}_{FE} = \frac{5.7 \times 10^{-12} \times 10^{4.52\phi^{-0.5}} A_e (\beta E)^{2.5}}{\phi^{1.75}} e^{(\frac{-k\phi^{1.5}}{\beta E})}$$

$$\frac{d(\log_{10} I_F / E^{2.5})}{d(1/E)} = -\frac{2.84 \times 10^9 \times \phi^{1.5}}{\beta}$$

Note that although these expressions are derived for ac fields, they do not show any frequency dependence. From the second equation, it is possible experimentally to obtain the value of β from the slope of emitted current measurements as a function of inverse electric field 1/E. Results are shown in Fig. 10 for the above seven-cavity S-band structure vs. successive dates of RF processing. The values of β are all greater than 50. This is very puzzling because calculations show that if the emitters were made of metal protrusions, as shown in Fig. 11, it could not exceed a value of 10. It indicates that there is another additional mechanism, impurity, dielectric or positive ion layer formation (not well understood yet) that enhances the value of β. But whatever the exact mechanism is, it seems to result in enhanced fields on the order of GV/m, huge instantaneous local currents and Joule heating, explosion of the protrusion with melting of the copper, formation of drops and surface damage [3].

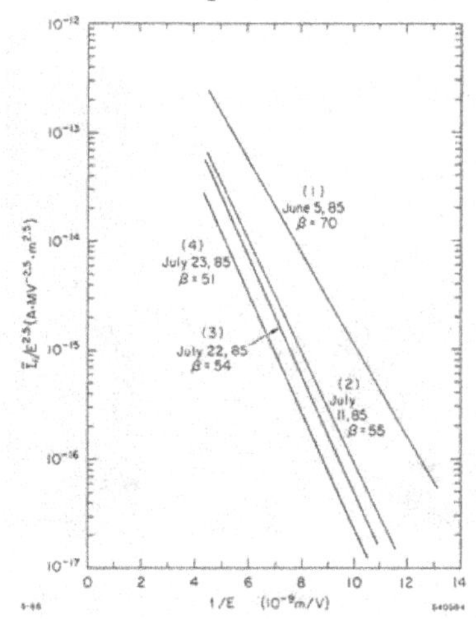

Fig. 10. Experimental measurement of the field enhancement β as a function of successive processing dates in the seven-cavity S-band structure.

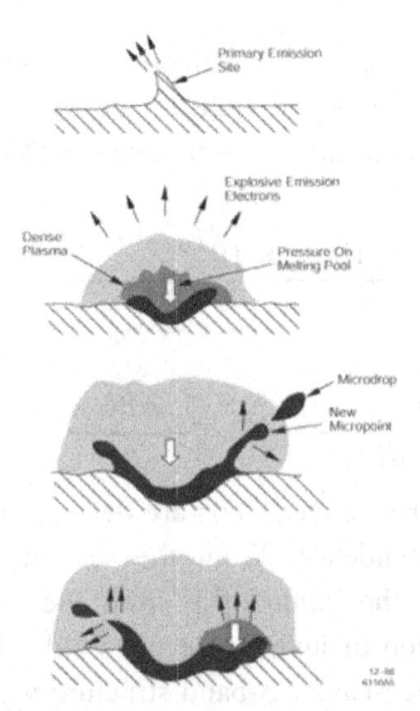

Fig. 11. Suggested model of breakdown resulting from field emission from protrusion followed by copper melting, drop formation and surface damage.

Note that since these measurements were done with a tunable RF source, it was possible to track the changes in the effective resonant frequency of the cavity without difficulty. As we will see later, this would not be possible with a practical multi-cavity accelerator structure where the detuning would not be the same in each cavity.

Fig. 12. Juwen Wang, myself, and Juwen's brother, Juwu visiting in my office at SLAC.

Fig. 13. Juwen Wang graduating from Stanford for his Ph.D in 1989.

4. Juwen Wang's Work since his Ph.D. in 1989

After getting his degree, Juwen became a regular SLAC researcher in the Accelerator Physics Department. His work led him to attend many international linear collider workshops as listed in Fig. 14, contribute to many large inter-laboratory collaborations, and play a major role in numerous new discoveries and results.

Accelerator Physics

Year	Workshop	Location
1988	LC88	SLAC
1990	LC90	KEK
1991	LC91	Protvino
1992	LC92	Garmisch
1993	LC93	SLAC
1995	LC95	KEK
1997	LC97	BINP, Zvenigorod
1999	LC99	INFN, Frascati
2002	LC02	SLAC
2004	1st ILC Workshop	KEK
2005	2nd ILC Workshop	Snowmass

Particle Physics

Year	Workshop	Location
1991	LCWS91	Saariselkä, Finland
1993	LCWS93	Waikoloa, HI
1995	LCWS95	Morioka-Appi, Japan
1999	LCWS99	Sitges, Barcelona, Spain
2000	LCWS00	Fermilab Batavia, IL USA
2002	LCWS02	Jeju, Korea
2004	LCWS04	Paris, France
2005	LCWS05	Stanford, USA
2006	LCWS06	Bangalore, India

3/9-12/06 G.A. Loew SLAC

Fig. 14. List of international linear collider workshops.

Much work on advanced linear collider accelerator structures was done between SLAC and KEKunder the leadership of Juwen Wang, and Toshi Higo with his colleagues, as shown in Fig. 15 below.

Fig. 15. Toshi Higo (right) with other KEK collaborators visiting SLAC.

Fig. 16. Ted Lavine, Chris Adolphsen and Juwen Wang in NLCTA at SLAC.

With SLAC colleagues Ted Lavine and Chris Adolphsen, Juwen did many interesting tests in the Next Linear Collider Test Accelerator (See Fig. 16 above), but a surprise lay ahead. As a 1.8 m long constant-gradient accelerator structure was processed to a gradient of about 40 MV/m, the first few cavities experienced severe breakdown, damage and unacceptable frequency detuning, as shown in Fig. 17 below.

Fig. 17. First few cavities in 1.8 m constant-gradient X-band accelerator structure showing severe damage after processing of gradient up to 40 MV/m.

The explanation was not immediately obvious but it was eventually realized that these long structures (1.8m) had excessively high group velocities at the input. Fast propagation of high input power led to unexpected breakdown damage in the front-end cavities, and unexpected pulse heating in the input couplers. The pulse heating temperature increase was shown to be as follows:

$$\Delta T = \frac{H_t^2}{\sigma\delta}\sqrt{\frac{t}{\pi\rho C_\varepsilon \kappa}}$$

where H_t is the tangential magnetic field, T_p is the RF pulse length, R_s is the surface resistivity, ρ is the RF conductivity, c is the specific heat and Kappa is the thermal conductivity.

As a result of this new understanding, shorter structures had to be designed with lower group velocity.This led to a change from 2π/3 to 5π/6 phase advance per cavity, optimized cell shape, and lower magnetic field in the coupler design.Structures had to be damped and detuned for Higher Order Modes (HOM), with acceptable ratio of iris diameter to RF wavelength, a/λ. These requirements in turn led to many mechanical complications. Figure 18 shows samples of cavity disks with four manifold holes, and Fig. 19 shows a cut-away section of the front-end of a round damped detuned X-band structure (RDDS) with HOM manifolds and couplers.

Fig. 18. Samples of individual cavity disks with four HOM manifold holes.

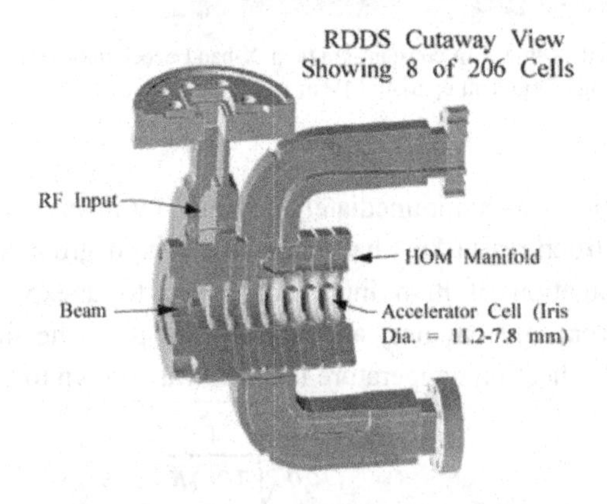

Fig. 19. Front-end of a round damped detuned X-band structure (RDDS) with HOM manifolds and couplers.

A complete 60 cm-long RDDS structure is shown in Fig. 20 below. Figure 21 shows statistics of the number of RF breakdowns for a total of 5 such structures after about 500 hours of operation, and 8 such structures after more than 1500 hours of operation at 60 Hz as a function of unloaded gradient (i.e. without beam). The acceptable rate of breakdown occurrences was set at 0.1 per hour for the structure. This shows that a 65 MV/m unloaded gradient could safely be reached at that time.

Fig. 20. Complete 60 cm-long X-band RDDS section.

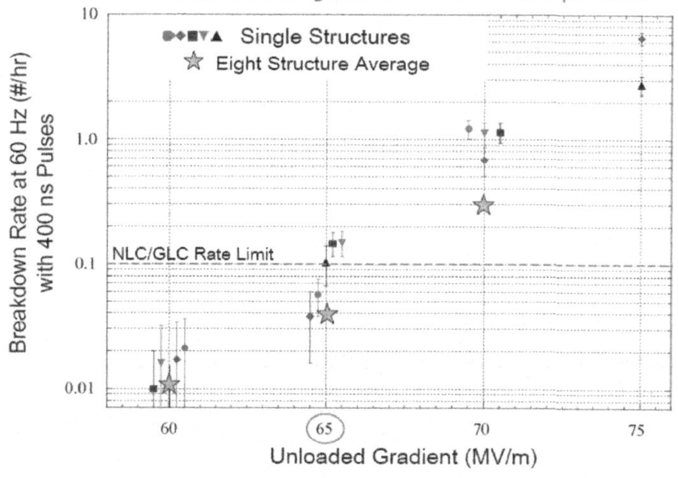

Fig. 21. High-gradient breakdown rate statistics for a total of 13 RDDS structures as a function of unloaded gradient.

But this was not the end of these observations. In a continuation of these studies, Sami Tantawi's group at SLAC later obtained new breakdown statistics withthree different cavities of slightly different shapes and different magnetic fields shown in Fig. 22 (also presented by Tantawi at this symposium).

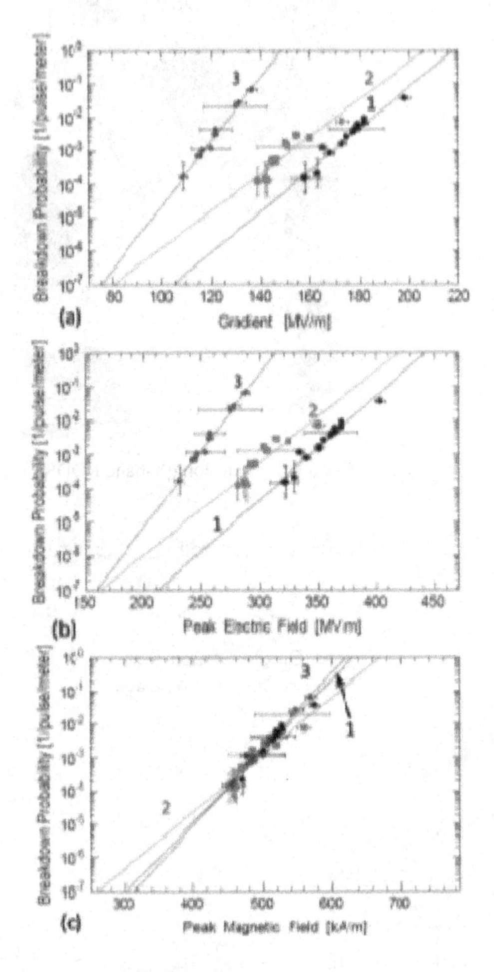

Fig. 22. Breakdown statistics for three different X-band cavity shapes as a function of (a) gradient, (b) peak electric field, and (c) peak magnetic field (reported by S. Tantawi at this symposium).

The interesting observation shown in Fig. 22c is that the breakdown rate was more closely correlated with the peak magnetic field than with the electric gradient (Fig. 22a) or the peak electric field (Fig. 22b). These results were not final at the time of this report but seemed to indicate that the initiation of the breakdown may be due to pulse heating in the disks, subsequently followed by field emission from the iris edges.

But Juwen's work didn't stop here. In the next two or three years, he helped build CERN/CLIC structures, the SLAC LCLS injector and X-band deflector structures shown below in Figs. 23 and 24.

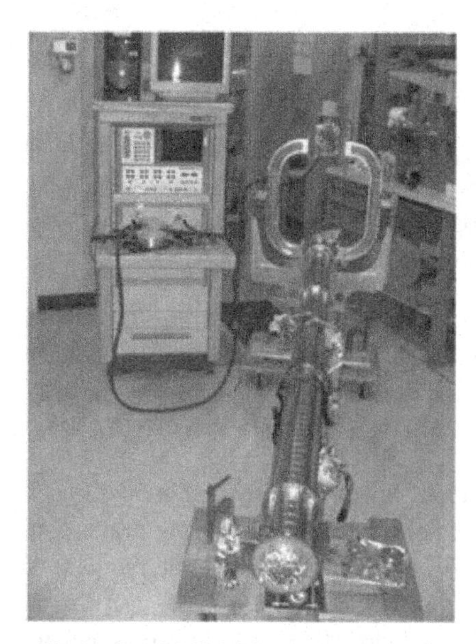

Fig. 23. SLAC LCLS injector structure.

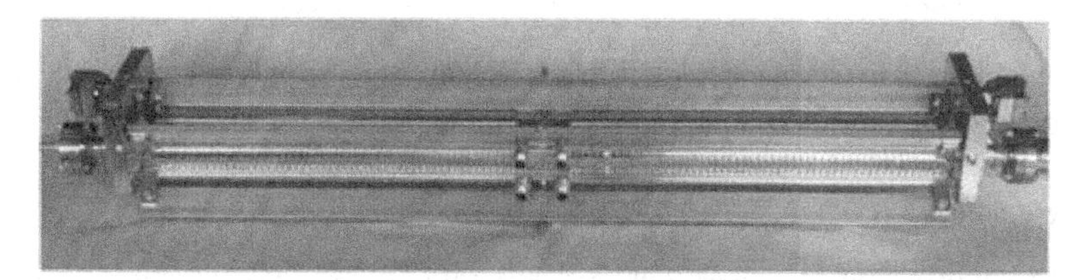

Fig. 24. X-band RF deflector structure.

Fig. 25. From left to right: Juwen, his daughter-in-law Jenny, his wife Xiuqin, his grandson Evan, and his son Frank Wang. Another grandson, Alan, was recently added to the family.

A happy picture of Juwen's family is shown in Fig. 25.

This concludes the written version of my oral report given on the occasion of Juwen Wang's 70th birthday celebration at Tsinghua University on May 19th, 2013.

As this report is being finalized, Juwen is engaged in work of a very creative design of a new SLED RF pulse compression system for the LCLS.

Acknowledgment

I wish to thank Juwen's son, Frank Wang, for supplying me with some biographical data and photographs of his father's life.

References

1. J.W. Wang, SLAC-Report-339.
2. R.H. Fowler and L. Nordheim, Proc. Roy. Soc. A119, 173–81, 1928.
3. G.A. Mesyats, IEEE Trans. Elec. Insul. Vol. EI-18 No.3 June 1983.

Research of RF High Gradient Physics and Its Application in Accelerator Structure Development*

Juwen Wang

SLAC National Accelerator Laboratory
Menlo Park, CA 94025, U.S.A.

The design of high-gradient room-temperature RF linear accelerators requires that special attention to be paid on several phenomena which limit their performance: electron field emission, dark current, RF breakdown, pulse heating, and cavity damage.

In this talk, the motivation, experimental and theoretical studies, and adventures in past more than 30 years at SLAC are introduced and reviewed. As most critical issues, these basic studies have greatly benefited the successful development for 100 MV/m high gradient accelerator structures applied in multi TeV linear colliders. Some important design practices and results as well as future work are also presented.

1. Special Acknowledgment

I feel specially honored to be given this opportunity to review some of our important research work, which we all dedicated on for many years. Words cannot describe how grateful to attend in this special symposium. I have been very fortunate to be benefited from so many wonderful people in my whole career, some of them are among the distinguished guests here: Greg Loew and Roger Miller were my advisers at US; Naiquan Liu, Weixie Gui, Yumin Hu, Yuzheng Lin and Dechun Tong were my teachers at China; Chuanxiang Tang, Zhentang Zhao, Huaibi Chen and Wenhui Huang are close friends supporting me for the collaboration work with Tsinghua and SINAP; Sami Tantawi, Derun Li, Nobu Toge, Toshi Higo, Walter Wuensch, Roger Jones, Qing Qin, Qiang Gu. Zhenghai Li and Liling Xiao and etc. are my dear colleagues. Also, I'd like to give my sincere thanks to Wei Gai, Jiaru Shi and the organizing committee for their great effort to organize this symposium.

---

*Work supported by U.S. Department of Energy, contract DE-AC02-76SF00515.

2. Introduction

The design of high-gradient room-temperature RF linear accelerators requires that special attention paid to several phenomena, which limit their performance:

- Electron field emission
- Dark current
- RF breakdown
- Pulse heating
- Cavity damage

All these phenomena are well defined and more institutions around the world are consistently make great efforts and slowly make continuous progress. As one of the pioneer researchers in this field, I'd like to provide some useful information and discussion in this occasion.

The outline of this paper is the followings:

- Contributions from Pioneer High gradient Studies [1].
- Progress in Recent Years.
- Applications in High Gradient Structure Development.

3. Contributions from Pioneer High Gradient Studies

In early 1980s, with the advent of the SLAC electron-positron linear collider (SLC) in 100 GeV center-of-mass energy range, research and development work on even higher energy machines of this type has started in several laboratories the United States, Europe, the Soviet Union and Japan. At SLAC, we set our research goals clearly in 1981 for an extensive research program: 1) in a fundamental way, to contribute to the understanding of the RF breakdown physics phenomenon, and 2) as an application, to determine the maximum electric field gradient that can be obtained and used technically safely in future e^+e^- linear colliders approaching the TeV energy range. Several important pioneering efforts in analysis theory, experimental methods and technologies were successfully developed, which now are still widely used in the high gradient accelerator research worldwide.

3.1. *Application of Fowler–Nordheim plots to analyze RF field emission and surface conditions*

The theory of DC Field Electron Emission (FEE) from ideal metal surface was obtained by Fowler and Nordheim with calculating the quantum mechanical tunneling of conduction electrons through a modified potential barrier at an ideal metal surface in an applied electric field. Furthermore, the Enhanced Field Emission (EFE) was categorized due to large variations in the microscopic surface field roughness called Geometric Field Enhancement as well as contaminations on the metal surfaces called Non-Metallic Field Emission Enhancement.

For the RF field in accelerator structures, we took the time-averaging for field and made some mathematic approximation treatments, the field-emission current for an alternating field can be expressed as:

$$\bar{I}_F = \frac{1}{T}\int_0^t I_F(t)dt$$

$$\bar{I}_F = \frac{5.7 \times 10^{-12} \times 10^{4.52\times\phi^{-0.5}} A_e(\beta E_0)^{2.5}}{\phi^{1.75}} \exp\left(-\frac{6.53 \times 10^9 \times \phi^{1.5}}{\beta E_0}\right)$$

where ϕ is the work function in *eV*, *Eo* is the amplitude of the macroscopic surface field in V/m, β is the enhancement factor and $\bar{I}_F$ is the average field emission current in Ampere on an emitting area Ae in m^2.

We can evaluate the surface condition to make measurement of field emission and obtain the field enhancement factor β from the slot of so called the Fowler-Nordheim Plot:

$$\frac{d(\log_{10} I_F/E^{2.5})}{d(1/E)} = -\frac{2.84 \times 10^9 \times \phi^{1.5}}{\beta}$$

Here is a nice story to recall. In the early stage of high gradient experiment on a SW S-band disc-loaded structure, nobody could realize and predict how strong the field emitted electron current could reach. The dark current was only picked up by a thin copper probe in the front of a downstream stainless steel sealing plate as shown in the Fig. 1 (right). As the RF power increased, the monitored temperature of the end–plate rose to 400°C and before the RF power could be turned down by me and Dr. Loew, the section vacuum already went to the air. As shown in the picture, here was a 1 mm diameter hole melted and punched by a well-focused, captured and accelerated (up to 10 MeV, few mA peak) electron dark current beam with 100 W average power, the exploded stainless steel was beautifully coated on the surfaces of several adjacent disks.

3.2. *Development of complete instrumentation for systematic studies on pre-breakdown and breakdown phenomena*

Through many years' experimental studies, we have cumulated experience to build complete instrumentations and tools to explore every important aspects of

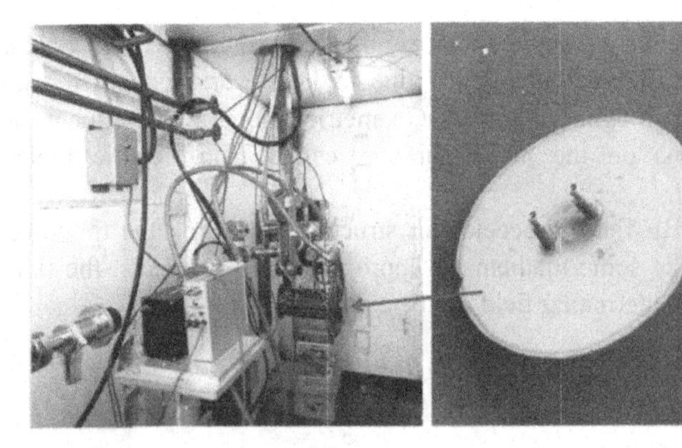

Fig. 1. Set-up of a seven cavity S-band Standing wave section (left) and a hole in its end plate punched by strong dark current (right).

pre-breakdown and breakdown phenomena. As shown in Fig. 2 and all attached plots, they typically include:

- Outgassing studies at high power RF: Residue Gas Analyzer (RGA) and Vacuum gauges.
- Radiation studies around structures: Collimated Scintillators plus Photo-multiplier tubes (PMT), Ion Chambers.
- Dark current studies: Profile monitors, Faraday Cups and Spectrometers.
- RF Breakdown studies: Instrumentation for Forward / Reflected RF waves.
- Surface analysis for before / after breakdown events: Scanning Electron Microscope (SEM), X-ray Photoelectron Spectroscopy (XPS) and Atomic Force Microscopy (AFM).

The above basic experimental methods and building blocks have become essential and widely used for any high gradient RF studies of the accelerator structures and components worldwide.

3.3. *Understanding of RF processing and breakdown mechanisms*

Based on the high gradient tests for many accelerator structures, we have gained much better understanding of RF processing and breakdown mechanism.

The RF processing is considered to be:

- Removal of adsorbed gases and surface dielectric impurities.
- Burning of micro-protrusions.
- Smoothing of surface roughness and pits.

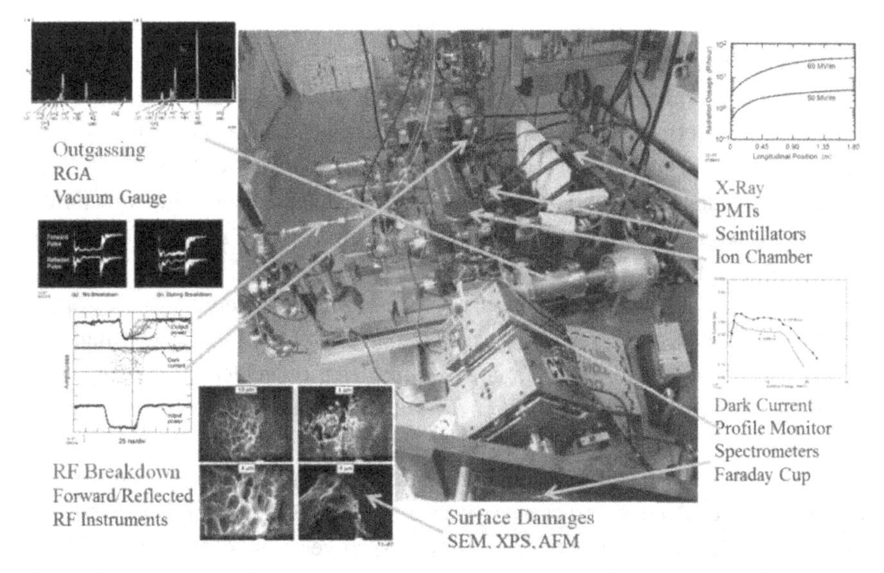

Fig. 2. Illustration of a typical setup for high gradient experiment: a 75 cm X-band TW accelerator section under high power test.

- Short RF power induced breakdowns, followed by protracted steady application RF power until field emitted current is decreased to lower asymptotic level.
- RF structure is protected by lowering power level due to either vacuum increase, bursts or large power reflection as a consequence of the RF breakdown events.
- With continuous RF processing from that slightly lowered level, the RF power can recover and keep increasing if the processing improves the structure condition. The repeated procedure will lead to the field emitted current decreased to lower asymptotic level and RF power increased to higher threshold.

The RF breakdown is considered to be Explosive Electron Emission companied with plasma formation and creation of surface craters. At a microscopic field of several GV/m, field emitted current density j_{FE} can approach 10^{11}A/m^2. At this level and above, the ri^2 Joule losses in an emitter can heat it in a few or tens of nanoseconds to the point of vaporizing some of its contents (gas, metal, etc.). This process can be beneficial by smoothing and cleaning the surface. However, the positive ions, produced by ionization of the neutral gas by the FE electrons, can form a sheath or plasma spot over the emitter, precipitating further increased FE, more ions, eventual explosion or meltdown, metal droplets, and expanding damage in the form of craters. Experimental evidence for these phenomena is corroborated by the well documented observation of visible sparks, instantaneous increase of the emitted current by a factor of ~40, detection of puffs of gaseous H, CH_4, H_2O, CO and CO_2, X-ray bursts, and collapse (within the RF pulse) of the structure fields, manifested by a reflection of the incident RF power and/or a sharp decrease in the transmitted field.

Experimental data obtained from early tests with a variety of single copper cavities and two short resonant accelerator sections [2] indicated that the maximum surface fields, Es, attainable after aggressive RF processing, i.e., where the incident power was driven up without concern of producing surface melting and craters, were ~310 MV/m at 2.856 GHz and 500 MV/m at 11.424 GHz. However, the resulting surface damage would have been unacceptable for a practical accelerator structure because the craters severely detuned the operating frequency of the cavities. In the last ten years or so, the drive to build reliable high-gradient structures for electron-positron colliders led to a very large number of room-temperature structure designs and tests, mostly at 11.424 GHz and some at around 30 GHz.

4. Recent Progresses

4.1. *Structures pulse heating and cavity damages*

Recent experiments cast at least some doubt on initiation of RF breakdown due to Strong field emission [3]. Although the entire mechanism is not yet understood, it appears that pulse heating, somewhere in the cavity where the magnetic field is high, may be the original cause. Somehow, this pulse heating may destabilize the cavity and then lead to FE and breakdown at the disk edges.

For pulsed RF accelerator structures made out of copper (or other metal), the magnetic fields and the corresponding electric currents in the neighborhood of irises or disk slot edges (used to damp higher-order modes), these magnetic fields and currents can theoretically cause initial heating exceeding 100 degrees C within a pulse T_p. A practical formula to estimate the temperature increase Delta T is given by:

$$\Delta T = \frac{H_t^2}{\sigma\delta}\sqrt{\frac{t}{\pi\rho C_\varepsilon \kappa}}$$

where Ht is the surface magnetic field, t is the time into the pulse, sigma is the electric conductivity, delta is the skin depth, ρ is the density, C_ε is the specific heat, and k is the thermal conductivity of the metal.

In extreme cases, this heating can induce thermal fatigue and thermal stresses beyond the elastic limit of the metal. After some millions of pulses, these are observed to produce surface roughness and cracking, and further enhanced pulsed heating. The threshold for the damage due to pulse heating is still unknown but the calculated temperature increase of over 100 degrees C for an initially undamaged surface gives an experimentally observed warning point not to be exceeded.

The observations made in tests of X-band (11.424GHz) accelerator structures have led designers to round off sharp edges in matching irises, coupling irises and disk slots to practical levels. For predicted temperature rises of less than 20 and 50 degrees C respectively, these improved structures have performed without any pulse heating damage. Note that these modifications have in no way affected the accelerating gradients of the structures.

4.2. *Basic physics research on short/compact accelerators*

In order to studies on some basic RF breakdown physics, many demountable shorts structures were designed, built and tested.

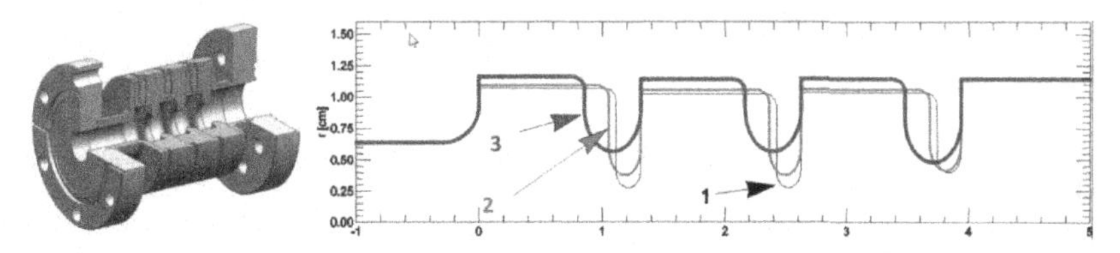

Fig. 3. Disk-loaded structures with different aperture and disk thickness.

Figure 3 shows three types of disk-loaded structures with respective a/λ values of 1) 0.105, 2) 0.143 and 3) 0.215. Surprisingly, as seen in the plots shown in Fig. 4 below, the probabilities of breakdown did not correlate with the gradient (a) or the peak electric field (b), but with the peak magnetic field in the cavity (c). This is an entirely new empirical result which does not give a complete mechanism of how the magnetic pulse heating somewhere in the cavity triggers an electric breakdown somewhere else.

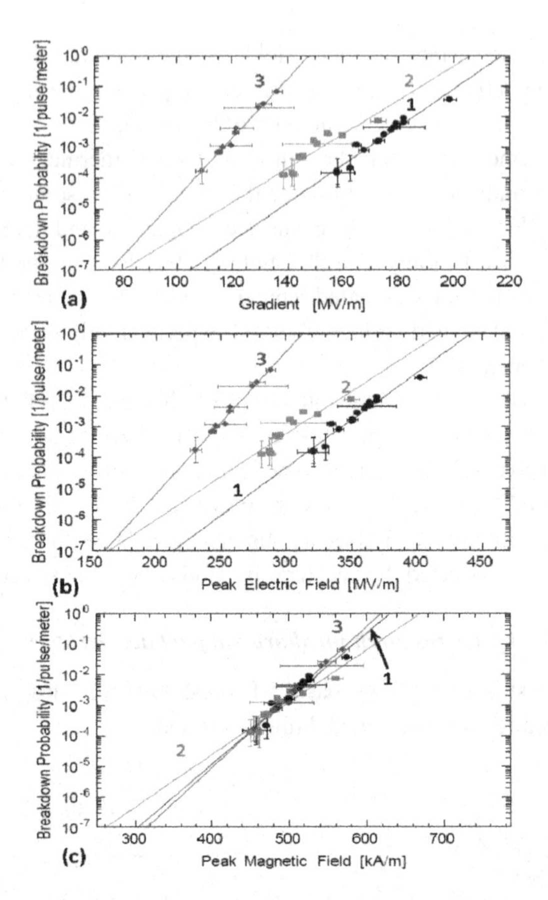

Fig. 4 Comparison of breakdown probabilities for three different structures, their respective a/λ values are given above.

5. Applications in High Gradient Structure Development

In recent years, all the knowledge gained from the basic high gradient research has greatly improved the accelerator structures design and their performances. Here we discuss some design issues, practical advanced high gradient accelerator examples and new manufacture technologies.

5.1. *Practical design issues*

Breakdown in these structures, once they have reached a desired level, results in two problems. The first is that during the pulses when breakdown occurs, the structure is unlikely to produce acceleration since power transmission is interrupted at some point. When this happens, the total energy of the accelerator will fluctuate in an unacceptable way. The second problem is that breakdown produces structure damages in the form of craters and crystal boundary cracks. If too many such damages are created, the frequency of the cavities where they occur will change. This detuning changes the phase velocity and reduces the net acceleration because the synchronism between the electron or positron bunches and the wave phase is affected.

The first problem is caused by imperfections in the copper (or other metal) surfaces, impurities, inadequate baking, incomplete thermal or RF processing, and by operating above thresholds which are not fully understood or predictable at this time. For example, all the X-band structures were treated with the following consecutive steps: accelerator cup surface light etching and cleaning, clean room assembly, wet and dry hydrogen firing and extended (two weeks) high temperature (650° C) vacuum baking.

The second problem is to reduce the RF breakdown damages. Apparently, we need to reduce the available RF power or energy for sustaining the RF breakdown. Therefore, it means lower input power P_{in}, shorter pulse width T_p, and lower group velocity V_g. Shorter structure (thereby increasing the total number of feeds per unit length) could need lower input power per structure. Selecting accelerating mode with higher phase advance has lower phase velocity; the filling time could be kept the similar with short structure length. By choosing proper aperture dimension, we can optimize the RF efficiency with much better high gradient performance. In summary, the preferable RF parameter options are Lower Vg, Short Length, and High Phase Advance.

5.2. *Special design issues*

- RF Couplers [4]

High gradient studies for several X-band accelerator structures have shown the heavy damages in the irises of RF input, which limited the operating gradient. The simulation indicated the pulse heating temperature increase above 100°C was reached at the sharp edges of the coupling irises due to high magnetic fields. One solution was to round and thicken iris horns as shown in Fig. 5b). To further reduce fields for higher power flow, a "waveguide coupler", which makes the waveguide itself an accelerating cell and power passes directly through an iris in the broad wall of the rectangular waveguide into a matching cell as shown in Fig. 5c).

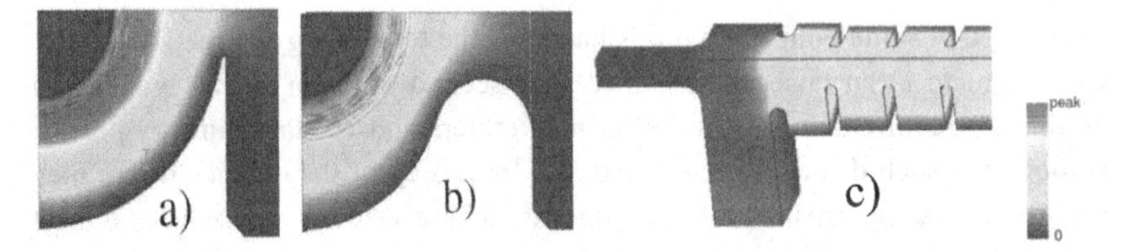

Fig. 5. Quarter cross-section of accelerator coupler for a) old fashion thin-iris coupler, b) fat-lip coupler with reduced surface magnetic field and c) waveguide coupler.

- Accelerator cavity optimizations: reduction of surface electrical and magnetic fields

5.3. *Design examples of complete high gradient linacs*

5.3.1. *GLC/NLC structures*

The NLC (Next Linear Collider) and GLC (Global Linear Collider) are $e^+ e^-$ linear collider proposals based on room-temperature accelerator technology [5]. There have been two major challenges in developing X-band (11.4 GHz) accelerator structures for the GLC/NLC. The first is to demonstrate stable, long-term operation at the high gradient (65 MV/m) that is required to optimize the machine cost. The second is to strongly suppress the beam induced long-range wakefields, which is required to achieve high luminosity. The development of high gradient structures has been a high priority in recent years. Nearly thirty X-band structures with various RF parameters, cavity shapes and coupler types have been fabricated and tested since 2000. This program has been a successful collaborative effort among groups at SLAC, KEK, FNAL and other labs. Figures 6 and 7 show the precision machined accelerator cups and prototype structures.

Fig. 6. Precision machined DDS accelerator cups for the GLC/NLC main linac structures.

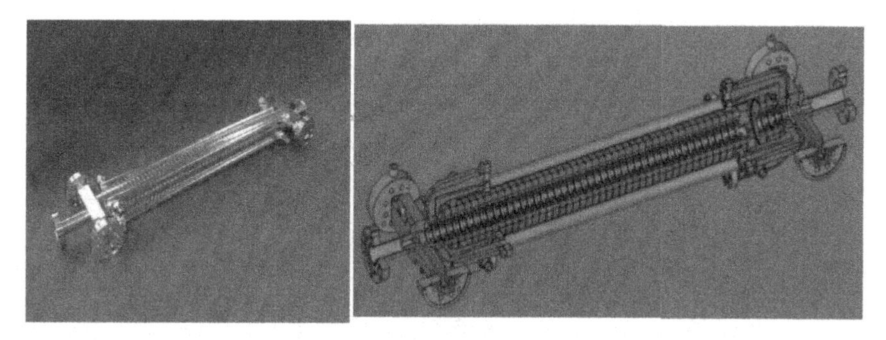

Fig. 7. Picture of the GLC/NLC prototype structure of H60VG4SL17 structure (left) and its cutaway view (right).

The first set of structures tested as part the high gradient program were low group velocity (0.03c ~ 0.05c) since the low-group velocity (downstream) portion of the 1.8m structures showed little damage. Although these structures proved much more robust, they have a smaller iris size ($a/\lambda = 0.13$) than required. Designing a low group velocity structure with $a/\lambda = 0.17 \sim 0.18$ is difficult since the simple solutions significantly lower the RF efficiency. The design that was adopted required increasing the phase advance per cell ($5\pi/6$ instead of $2\pi/3$) and using thicker irises to maintain a relatively high shunt impedance with the larger iris size. In order to maintain an optimal filling time, the structure length was scaled from the earlier 1.8m structures, which had group velocities of 0.12 c ~ 0.03 c, to a 0.6 m length with group velocities of 0.04 c ~ 0.01 c.

Figure 8 is a summary of many experiments at X-band at SLAC. The recipe for successful processing is to carefully process each structure to an asymptotic field level above which no more improvement is achieved within a practical time, but below the level where damage becomes inevitable. Another criterion which must be met is that for a given collider design, the number of allowable pulses lost per unit time due to breakdown is limited. Typically, for the NLC structures discussed here, this allowable number is below the green dotted horizontal line. Note that the breakdown times within each pulse were observed to be uniformly distributed. Figure 8 shows the average breakdown rate as a function of accelerating gradient after 500 and 1500 hours of processing respectively.

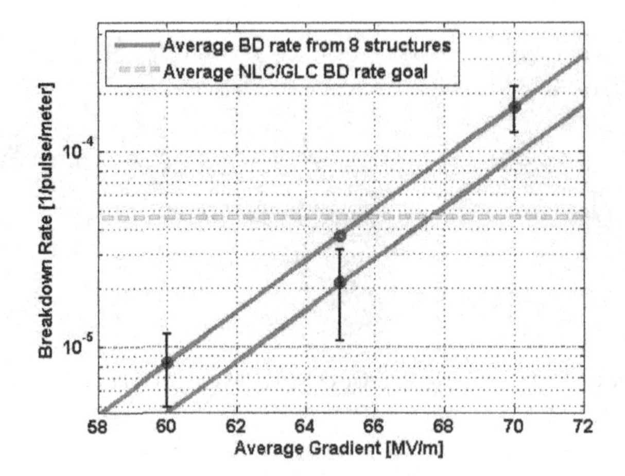

Fig. 8. Average breakdown rates for a series of NLC test structures as a function of accelerating gradient after 500 hours (upper line) and 1500 hours (lower line) of RF processing.

5.3.2. *CLIC structures*

An international collaboration on high gradient X-band accelerator structure development, which has been mainly lead by CERN, SLAC and KEK was started in 2007. Significant progress over the past few years has been made towards demonstrating the acceleration gradient of 100 MV/m with the nominal pulse width of 240 ns and a breakdown rate of a few 10^{-7}/pulse/m [6]. More than twenty prototype structures without and with the heavy damping features were designed, fabricated and evaluated their high gradient performance. Figures 9 and 10 show the pictures of T18 structure without damping features and TD24 structure with heavy damping feature. These structures were fabricated using the technology developed from 1994 to 2004 for the GLC/NLC linear collider initiative.

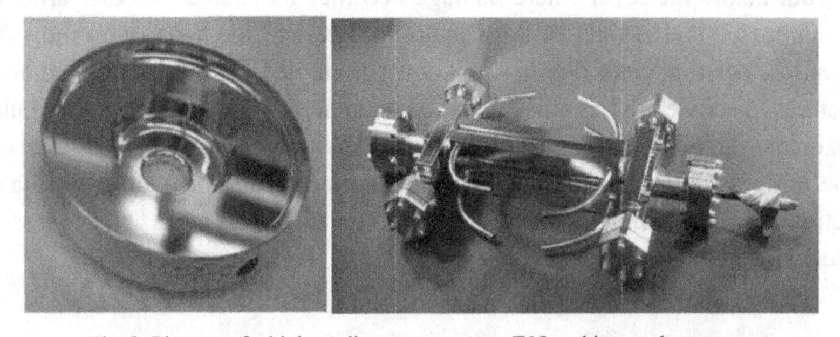

Fig. 9. Pictures of a high gradient test structure T18 and its accelerator cup.

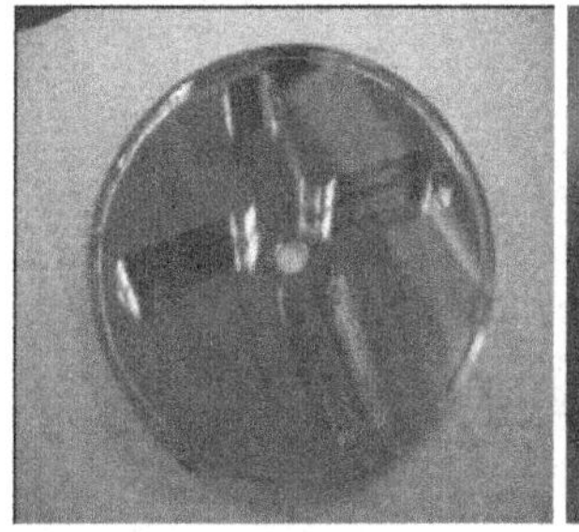

Fig. 10. Pictures of a prototype structure TD24 for the CERN CLIC main linac and its accelerator cup.

During the past few years, many structure tests at NLCTA (SLAC), Nextef (KEK) and CTF (CERN) were performed for more than twenty CLIC structures with four basic groups: T18, TD18 and T24, TD24 structures, where the D means with heavy damping feature. 18 or 24 means the number of regular accelerator cups in $2\pi/3$ mode. The goal is to use a lower group velocity, smaller aperture and shorter pulse length (230 ns vs 400 ns for NLC/GLC) - all of which have generally yielded higher gradients - to achieve at least 100 MV/m unloaded gradients reliably (compared to 65 MV/m for the NLC/GLC structures). However, the smaller average a/λ (13% vs 17-18% for NLC/GLC) produces stronger transverse wakefields that require tighter structure alignment tolerances and more aggressive long range wakefield control, in particular, having local HOM damping with Q's of order 10. Figure 11 summarizes the encouraging results to confirm the design goal could be reached for RF breakdown rate lower than 3×10^{-7}/pulse/m at average 100 MeV/m accelerating gradient.

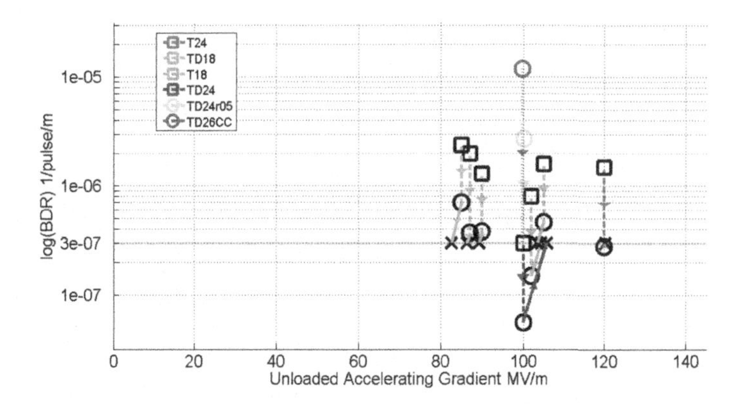

Fig. 11. High gradient performance summary for CLIC test structures.

References

1. J. W. Wang, SLAC-Report-339, July 1989.
2. G. A. Loew and J. W. Wang, XIII International Symposium on Discharges and Electrical Insulation in Vacuum, Paris, SLAC-PUB-4647, May 1988.
3. V. Dolgashev, S. Tantawi, Y. Higashi, B. Spataro, Applied Physics Letters, Vol. 97, 2010.
4. C. Nantista, S. Tantawi, V. Dolgashev, Phys. Rev. ST Accel. Beams 7, SLAC-PUB-10575, July 2004.
5. Juwen Wang and Toshiyasu Higo, ICFA Beam Dynamics Newsletter *No. 32,* SLAC-PUB-10370, Feb. 2004.
6. Walter Wuensch, CLIC Workshop, 2013.

Advances in the Understanding of the Physical Processes of Vacuum Breakdown

Walter Wuensch

CERN, Geneva, Switzerland

Advances in the fundamental understanding of vacuum breakdown achieved during the program to develop high-gradient accelerating structures for a TeV-range linear collider project are described.

1. Introduction

One of the main parameters that determines the performance of CLIC, a TeV-range e+e- linear collider [1], is the accelerating gradient achievable in its normal-conducting, rf (radio-frequency) accelerating structures. This gradient is limited by a number of effects, but the dominant one is vacuum breakdown. As a consequence, significant effort has been invested into understanding vacuum breakdown, both into the aspects specific to high-gradient rf accelerating structures but also, more generally, into the fundamental physics of the process. Advances made by an international collaboration, roughly centered on the CLIC study, to understand vacuum breakdown are reviewed in this report.

CLIC accelerating structures operate at X-Band, specifically 11.994 GHz, and are made from copper. The operating accelerating gradient is 100 MV/m which results in a peak surface electric field of approximately 250 MV/m. Numerous different versions of prototype structures have now successfully operated at 100 MV/m accelerating gradient [2, 3]. The aforementioned references [1–3] and references contained therein, contain detailed descriptions of the structure designs, fabrication methods and test conditions.

The effort to understand the basic physical processes of vacuum breakdown, and the implementation of numerous ideas based on this understanding, has contributed directly to the success in achieving the very ambitious accelerating gradient target of 100 MV/m. A summary of accelerating gradients achieved during testing is shown in Fig. 1. In the opinion of the author, the achieved high gradients give strong evidence of the validity of many of the new insights that were acquired during in the investigation. Some of the insights relate to specificities required for a high-energy linear collider but many apply quite broadly to vacuum breakdown in general. Also the high-gradient accelerator technology, developed for linear colliders, may be directly useful for other linac applications such as free electron lasers, Compton light sources and for medical linacs.

This report first covers the characteristics of vacuum breakdown in high-gradient rf accelerating structures and outlines how these relate to linac performance requirements. This gives the context for the breakdown study which has been driven by the linear collider project. The report then covers the basic physics of breakdown especially addressing the distinct experimental signatures which are observed in high-gradient rf. This section covers the importance of material microstructure, field emission, plasma build up and the interaction with external driving circuits.

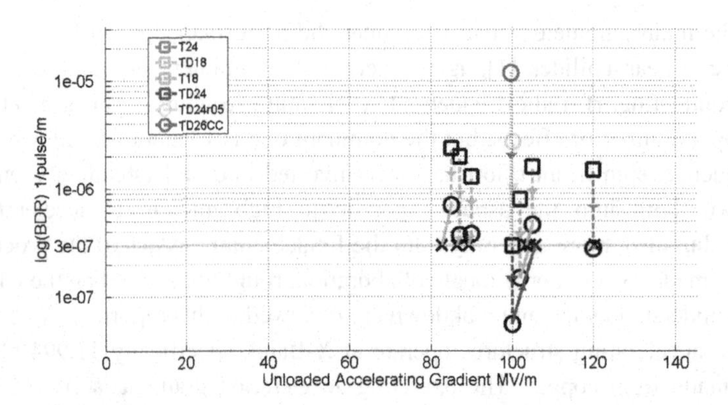

Fig. 1. Summary of performances of various versions of CLIC accelerating structure. CLIC operation requires a low breakdown rate, y-axis, as explained below.

2. Characteristics of Vacuum Breakdown in High-Gradient RF

The CLIC linac, like most normal conducting linacs, operates in a pulsed mode. The CLIC rf pulse length is about 200 ns and the repetition rate is 50 Hz. One of the key measures of the performance of accelerating structure is the breakdown rate at the operating gradient. The breakdown rate is the probability that a breakdown occurs on any given pulse, and is determined experimentally by dividing the number of breakdowns by the total number of rf pulses during an operating period. The CLIC specification is very demanding, requiring a breakdown rate less than $O(10^{-7})$ [1].

Breakdown rate (*BDR*) is observed to depend very strongly on the field level in the structure [4], and can be approximated by $BDR \alpha E^{30}$ [5]. Experimental data of *BDR* vs. *E* is shown in Fig. 2. The *E* shown on the abscissa is accelerating gradient, but could be any field in the structure since the electromagnetic fields all scale proportionally, giving the same exponent. Most fully conditioned structures are best fit with an exponent of around 30, although this tends to lower values during the early stages of structure conditioning. Fitting the breakdown rate to a power of field inside the structure is not based on a physical model but it does allow a simple, single parameter, quantitative comparison between structures. Determining what determines the statistical properties and dependencies of breakdown rate has been an important priority for this study, both because our particular application requires operation in a low breakdown rate regime and because the dependence represents a distinctive experimental signal that has the potential to give deep insight in the breakdown process.

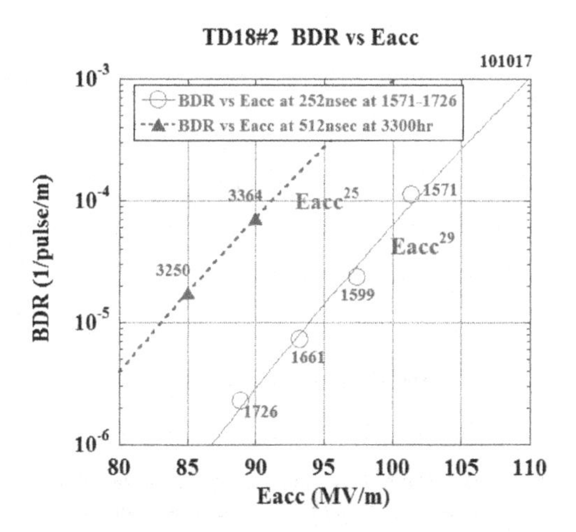

Fig. 2. Experimental data showing the strong dependence of breakdown rate on field of a CLIC high-gradient test structure [6]. The structure was a so-called TD18 tested at the NEXTEF test stand at KEK.

The breakdown rate is also a strong function of the rf pulse length. This observed dependence is typically about $BDR \alpha \tau^{5}$. Experimental data is shown in Fig. 3. In addition to its relevance as an experimental signal for understanding breakdown, this dependence of breakdown rate on field is important since the efficiency of the CLIC linac is strongly dependent on the rf pulse length. A longer rf pulse length allows a longer bunch train to be accelerated.

An important insight into the mechanism that drives the statistics of breakdown can be deduced by considering the pulse length dependence and the distribution of the time inside pulses when breakdowns occur. If the chance of a breakdown was entirely independent on every pulse, the probability distribution would have to increase strongly at the end of the pulse to give the observed pulse length dependence. This is because the structure does not know at the beginning of a pulse how long the pulse will become, so the τ^5 increase in breakdown probability would have to come from the end of the pulse. However, experiments show that the distribution of breakdown time within a pulse is essentially flat, Fig. 4. This means that there is a memory effect or evolution of the structure surface over a certain number of pulses. The early part of a pulse can know how long the pulses are because it knows how long previous pulses have been. A possible mechanism to explain such an effect is a local fatigue process near the field emitter sites, which we will return to later.

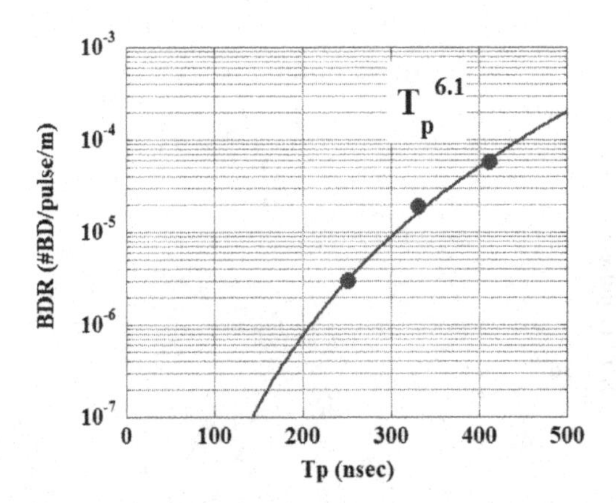

Fig. 3. Experimental data of breakdown rate as a function of rf pulse length from a TD24R05 tested at NEXTEF [3]. This particular data shows a fit of τ^6, a more typical dependence is τ^5.

A schematic layout of a typical high-gradient experiment is shown in Fig. 5. Incident, transmitted and reflected signals on pulses without and with breakdown taken during the test of a CLIC structure are shown in Figs. 6a and 6b. Three distinct phases typically appear in pulses in which a breakdown occurs. The first phase is the unperturbed pulse before the breakdown. When the breakdown begins, the transmitted pulse drops to near zero transmission in a time of about 10 ns. A strong reflection eventually develops, rising up to nearly

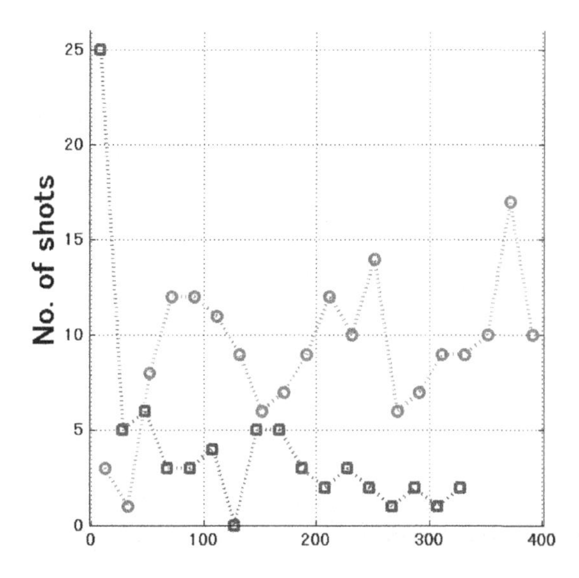

Fig. 4. Histogram of experimental data of breakdown time within a pulse from the same structure as Fig. 2. The red points show a basically flat distribution of breakdown times with a pulse, and clearly does not rising towards the end. The red points are those breakdowns which are preceded by at least one pulse which had no breakdown, so represents the intrinsic statistics of breakdown. The blue points are from breakdowns which occur on those pulses which have been immediately preceded by another breakdown. They show the effect of a breakdown, which may be to create a debris which must be cleaned up. The data was taken with 400 ns pulse length.

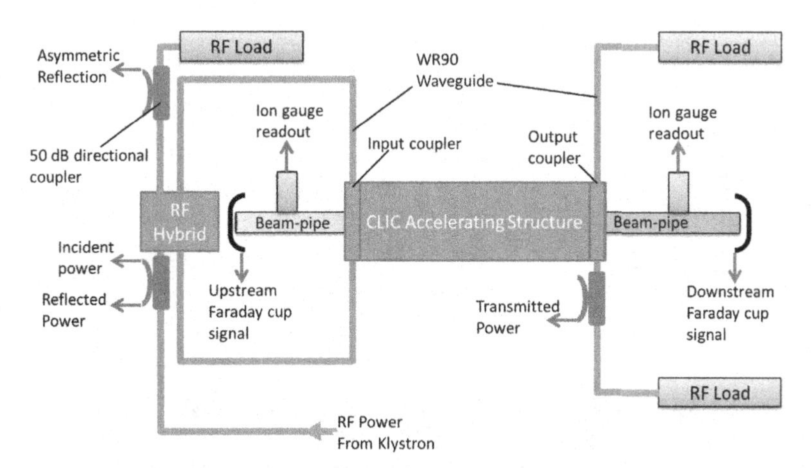

Fig. 5. Typical set-up of a high-gradient rf experiment. Drawing taken from [7].

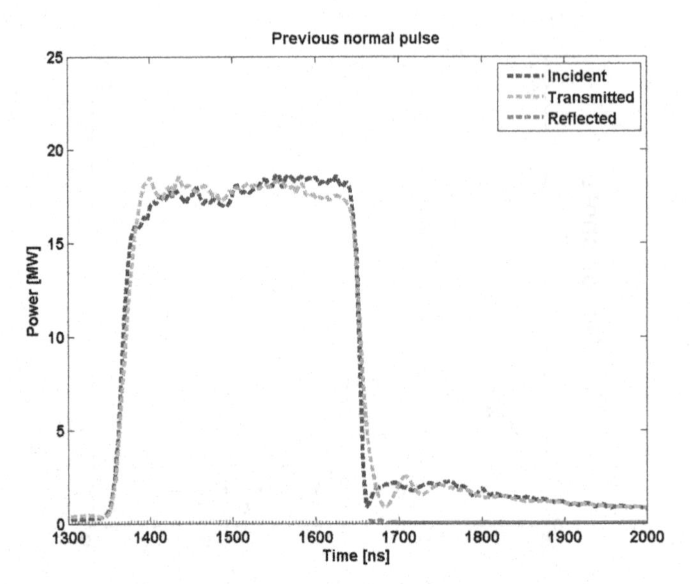

Fig. 6a. Incident (blue), transmitted (green) and reflected (red) signals on a pulse without breakdown [4, 6]. The rf pulse traveling through the structure has been normalized for attenuation due to ohmic losses. The data are from a TD26CC structure tested at Xbox-1 at CERN.

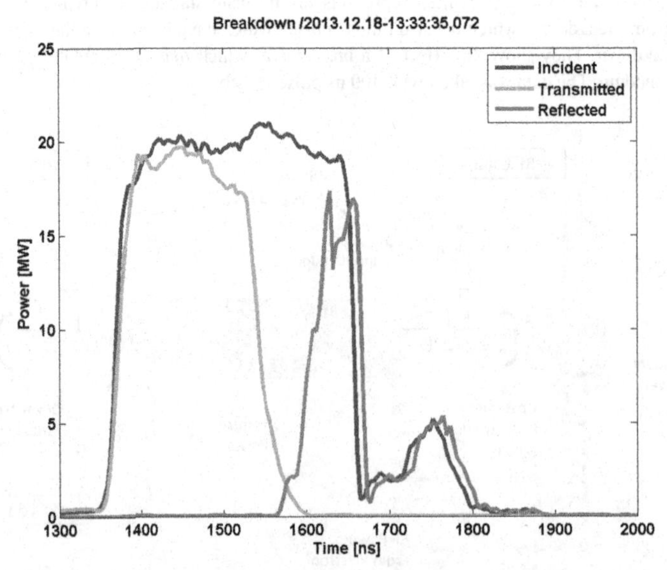

Fig. 6b. Incident, transmitted and reflected signals on a pulse during which a breakdown occurred. The signature is the abrupt cut-off of transmitted power, the drop in the green curve. Sometime later the incoming power is reflected from the breakdown.

full reflection. When there is no breakdown the forward power is equal to the transmitted plus reflected power (usually less than 1%), once cavity ohmic losses are taken into account. The common jargon says that there is no 'missing power' or 'missing energy'. However when a breakdown occurs, significant missing power is observed in the time between the transmission fall off and the reflection rise. The total missing energy of the pulse can exceed 50%.

Another important experimental observable in high-gradient rf experiments is the emission of currents from the structure. These currents are detected in Faraday cups that capture current emitted from the beam pipes of the structures (that is along the same path through the structure as the accelerated beam). Experimental data of current emission are shown in Figs. 7a and 7b. During rf pulses without breakdown this emission has a level below a milli-ampere. When a breakdown occurs, however, this level rises to many amperes. It should be emphasized that the total currents generated by the breakdown will be significantly higher since only a small fractional solid angle of current can escape the structure.

Prototype accelerating structures for CLIC are prepared according to a baseline procedure which has evolved over many years [8,9] for which the main

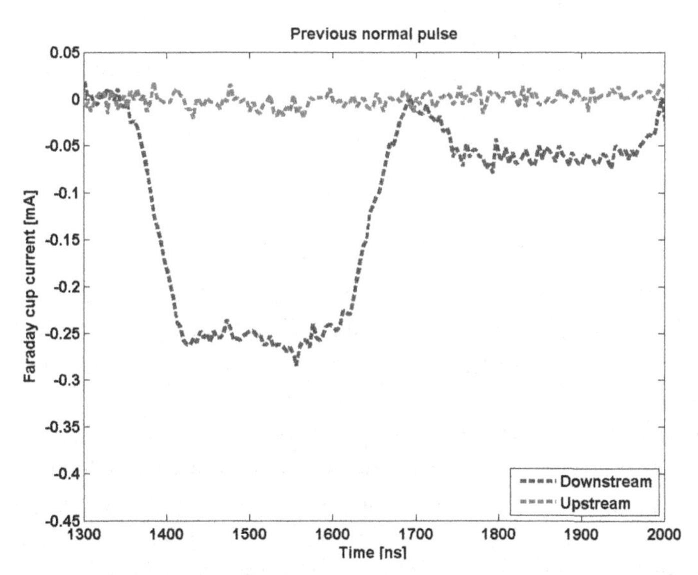

Fig. 7a. Current emission, so-called dark current, from a structure on pulse without breakdown.

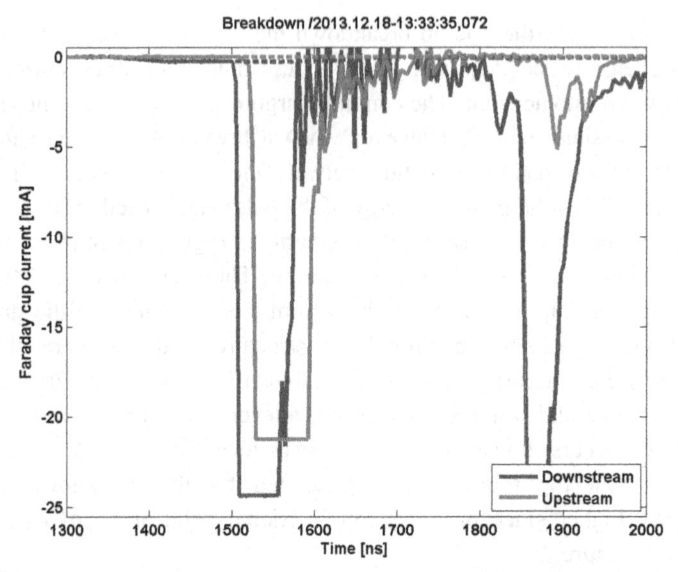

Fig. 7b. Current emission from the structure on pulses with breakdown. During structure testing dark current is usually monitored. This typically results in saturated signals during breakdown.

features and steps are:

- The raw material is OFE copper.
- Final machining is made with single crystal diamond tools on micron precision lathes and mills.
- Assembly in a clean room.
- Bonding at 1030°C in a 1 atm hydrogen atmosphere.
- Vacuum bake-out at 650°C.
- Storage (and installation) under dry nitrogen (flow).

Despite the very elaborate preparation procedure, the structures still require an extensive rf conditioning period. Many weeks of operation at 50 Hz are required to bring the structure up to operating parameters. The operational history of an accelerating structure is shown in Fig. 8.

New structures are first run at low power level and with short pulse length. Initially there is a short period, of the order of days, during which the structure out-gasses during stable operation (without breakdown) and as the power is raised, the first breakdowns occur at low field level. These initial breakdowns are associated with large vacuum bursts. Gradually the field and pulse length can be increased. Breakdowns occur intermixed with stable operation, but

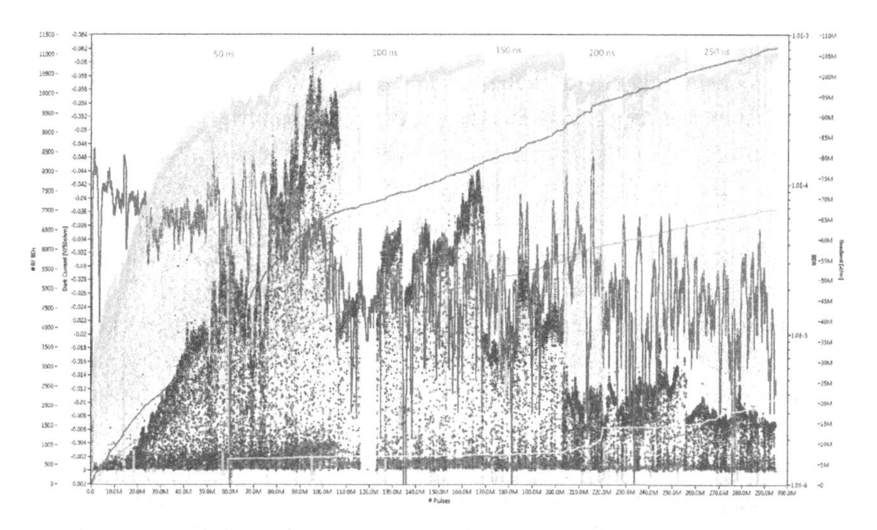

Fig. 8. Operational history of the conditioning process of an accelerating structure a CLIC TD26CC [6, 2]. The green trace is accelerating gradient (right hand axis). The red curve is the total number of accumulated breakdowns (left-most axis) and the purple line its derivative. The structure is conditioned with constant breakdown rate. The blue data is dark current level which can be seen to decrease with conditioning (middle axis on left side).

overall the performance of a structure will improve. Vacuum activity, both in normal operation and from breakdown, decreases significantly during this initial period.

The strong vacuum activity and breakdowns with strong vacuum bursts during the early stages of conditioning seems to be removal of, and triggered by surface contaminants such as particles and oxide layers. After this phase, it seems that the surface has become quite clean. Breakdowns then occur due to the imperfections and dynamics of the metal surface itself. The breakdown studies described in this report concentrates on these later stage breakdowns, and consequently focuses on those breakdowns which are initiated by the material atomic and micro-structure of a metal surface itself.

In order to minimize conditioning time without compromising ultimate performance, the strategy that has been adopted by the CLIC study is to condition with short pulses, typically 50 ns, go somewhat above the nominal gradient, expand the pulse length, go somewhat above the nominal gradient again and so on (many other rf applications use the same basic strategy). The surface treatment process made during fabrication is elaborate and the conditioning period can take many weeks — consequently preparing the surface for high-gradient contributes significantly to the cost of a completed structure. A strong practical motivation for a basic study of breakdown is to be able to refine the fabrication technique in a targeted, intelligent, manner. This could reduce fabrication time, cost and risk. Additionally, the conditioning strategy currently under use has been developed through trial and error. For example no set of identical structures has been conditioning in differing ways, so no statistically validated comparison exists. Here again a detailed understanding of the breakdown process can help guide development in a strategic way.

Another significant experimental observation from vacuum breakdown in radio frequency structures is that the performance of a structure with a particular geometry is limited by a non-simple function of the fields inside the structure. For example, the maximum surface electric field one finds in a fully conditioned accelerating structure, running at a specified rf pulse length and breakdown rate, depends very strongly on the geometry of that structure, even given the same preparation techniques. That is, the peak surface electric field does not provide the limitation in high field operations as one might expect from extrapolating from dc experiments. Broadly speaking, the dependency one sees is that a structure with a low power flow will sustain a higher surface electric field than a structure with high power flow [10]. This has led to the introduction of two quantities $P/\lambda C$ and S_c [4,11] that predict at which field a structure will operate – which then allows the operating accelerating gradient to be determined. The two quantities are based on a broad analysis of the observed performances of many structures, along with general physical considerations. The quantities appear to be rather accurate. Indeed, implementing the quantities from one generation of CLIC test structure to the next resulted in an improved high-gradient performance as shown in Table 1. Pairs of both the structures in the table were made with the same fabrication and surface treatment procedure, and were tested by the same teams, giving some level of assurance of reproducibility.

Table 1. Comparison of peak field values for two generations of test structures: T18 and T24. Both structures were tested at the NEXTEF facility at KEK. The data are given for a pulse length of 250 ns and a breakdown rate of $1\text{x}10^{-6}$ b/p/m.

	Accelerating gradient [MV/m]	Surface electric field [MV/m]	Peak P/λC [W/ μm2]	Peak Sc [W/ μm2]
T18	103	238	0.090	4.69
T24	118	243	0.106	4.77

In addition to their direct application in raising accelerating gradient, the scaling laws $P/\lambda C$ and S_c are used in the overall optimization of linear accelerators for construction cost and power consumption. A better foundation for the scaling laws, and possibly a revision, would allow a better optimization of the accelerating structure and facility design to be made. A quantitative understanding of the physical process which underlie vacuum breakdown, and in particular rf breakdown, might allow an even more accurate prediction of gradient as a function of geometry to be made. Some new directions are described in the last section of this report.

3. Breakdown Physics

CLIC accelerating structures are constructed from copper disks machined on ultra-high-precision lathes with a resulting surface roughness of below 10 nm. After high-gradient testing, structures are often cut open and inspected with optical and scanning electron microscopes. The craters seen in the rf test look very similar looking breakdown 'craters' are observed from sparks in the CERN dc spark system [12,13] as shown in Figs. 9a and 9b.

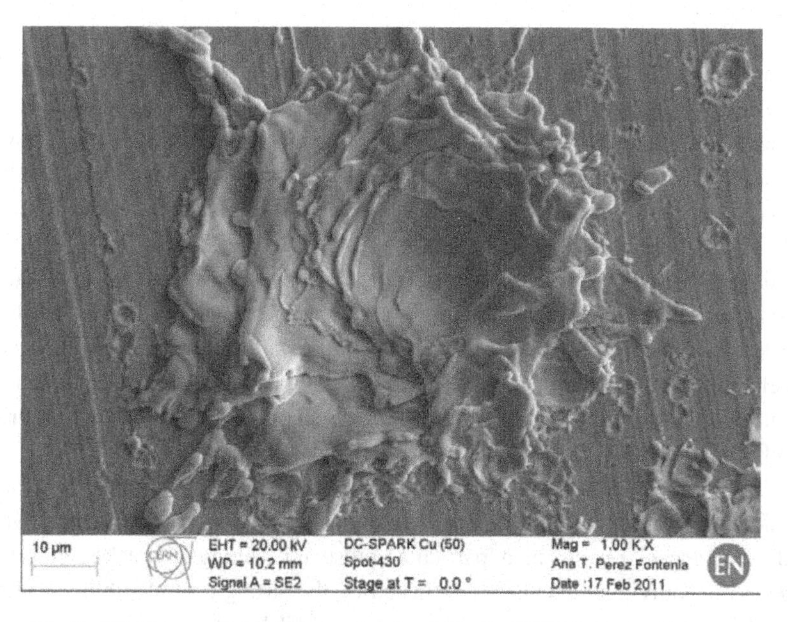

Fig. 9a. SEM image of a breakdown crater formed in the CERN dc spark system [14].

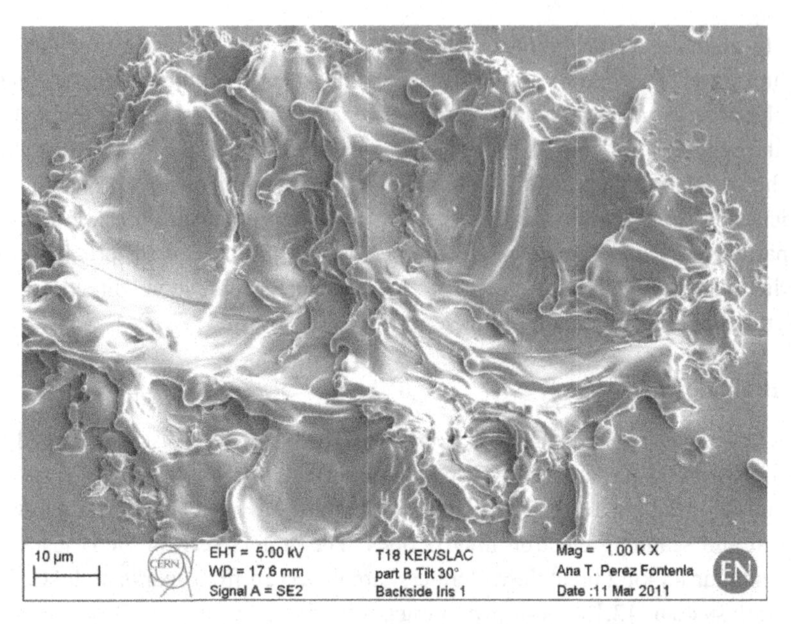

Fig. 9b. SEM image of a breakdown crater formed in an X-band accelerating structure [14].

Typically a few thousand breakdowns, defined according to the rf and faraday cup signals described above, are recorded during an rf structure test. Very roughly, a factor ten more craters are observed on a structure surface after it is cut open than are recorded by the diagnostic systems during high-power operation, indicating that there are undetected breakdowns and/or secondary breakdowns during the detected events. As we will discuss below, the estimated size of the features which trigger the breakdown is of the order of 10 nm. Assuming that the observed craters form around the area where the breakdown starts, the fraction of an accelerating structure surface which directly leads to breakdown during a typical test is below 10^{-8} (the high field region of an X-band accelerating structure is of the order of a cm^2). This is a very small number. In addition, structures are run at a breakdown rate of about 10^{-5} during conditioning and 10^{-7} during operation.

These simple considerations show that a very small fraction of a surface breaks down extremely rarely. What makes certain places on the surface weak and what aspect results in a particular pulse unleashing a breakdown? This rareness of individual events is contrasted by the strong reproducibility of field level from structure to structure. A summary of CLIC high-gradient test results was shown in Fig. 1. One can see individual structure types show performances that are the same to within a few percent, as are the high-power scaling limits summarized in Table 1.

The rareness of breakdown combined with the repeatability from sample to sample, plus the intra-pulse breakdown distribution combined with the observed τ^5 pulse length dependence of breakdown rate indicate, in the opinion of the author, that the features which lead to breakdown *evolve* to a characteristic feature type on the structure surface. Furthermore, for reasons which will be elaborated below, there is a strong indication that this evolution occurs through the dynamics of *dislocations* at or near the material surface.

The dark currents shown in Fig. 7 are observed in many other high-voltage applications, often referred to as field emission currents. The field enhancement factors β in measured CLIC structures are typically 30-60. This discussion of breakdown will continue with the common assumption [15], that these enhance emission sites are the locations at which breakdown can occur. Indeed experimental evidence of a threshold local electric field, βE value, has been obtained in the test in the CERN dc spark system [13], which supports this assumption.

What is the nature of these sites? In many high-voltage applications, breakdowns occur due to contaminants such as dust or oxides. These indeed seem to play a role in breakdowns early in the conditioning process of high-gradient accelerating structures, but the strong reproducibility of performance after conditioning, only a few percent as shown in Table 2, between structures argues against this being relevant in late stage breakdowns. In addition, a strong material dependence of threshold electric field has been observed in measurements in the dc spark system, Fig. 11. The material dependence indicates that the breakdowns which limit the ultimate gradient are features of the material itself.

A very compelling explanation for the ordering of the materials in Fig. 10 has been made in [20], which is that the crystal structure of the material determines the surface-field limit. Experimentally, face-centered cubic structures tolerate the lowest gradient, followed by body-centered cubic and ultimately hexagonal. This is consistent with the ordering of dislocation mobility, from highest in face-center to lowest in hexagonal. It is reasonable that higher dislocation mobility results in lower threshold surface field since features leading to breakdown become easier to form.

The common explanation for increased field emission current compared to that expected from the Fowler-Nordheim equation [21] applied to an ideal

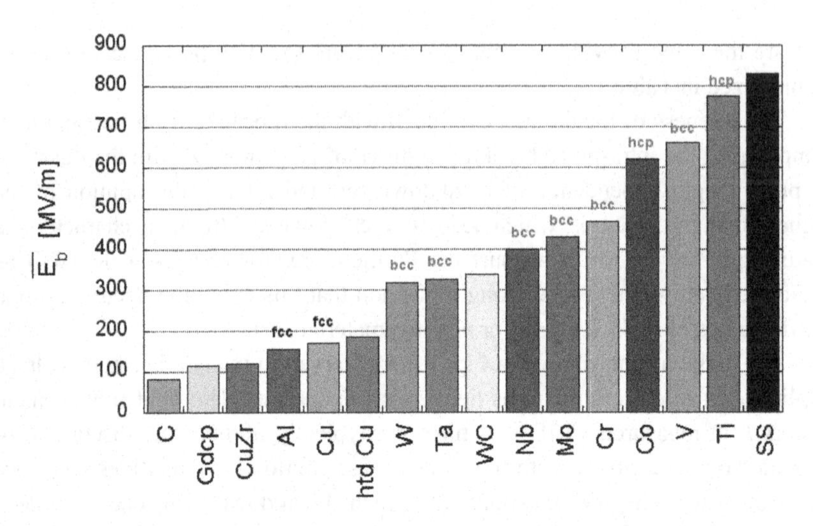

Fig. 10. Material dependence of threshold electric field from the CERN dc spark system [19].

surface is a local *geometrical* field enhancement factor (although enhanced emission is thought to come from dielectric dust particles in some applications). Our group has never independently observed, through SEM images for example, any features with the necessary aspect ratio to produce such a field enhancement on any of the surfaces we have analyzed, and this down to dimensions below a few hundred nm. It is interesting to note that geometrical surface features with dimensions of a few nm are thermodynamically unstable, and will disappear within a few seconds [22]. This limits the breakdown site size to between a few and a few hundred nm.

An additional contribution to field emission may be that a real surface is likely to have areas of locally reduced work function. The different crystal planes of copper have work functions which range from 4.48 to 4.98 eV [23]. Taking this idea further, interruptions in regular crystal structure, such as those caused by a dislocation at the crystal surface, should result in a locally lower work function compared to the bulk value. Experimental evidence for local areas of reduced work function has been reported in [24]. In this experiment, photo-emission from a copper surface using a 3.1 eV laser, that is less than the bulk work function, was observed. The total surface area of reduced field emission is small, and photo-emission is not enhanced exponentially like field emission, so the total current from the effect is small and the experiment is difficult to make. Studies to directly compute the effect of irregular crystal structure on work function are being made [25]. A first result from this study is that the effect of a single add-atom is to lower the work function of a copper surface by about 6%. This is a small contribution to a field enhancement factor of 50, but the effect is likely to be bigger for more complex structures. However the so-call DTF (Density Function Theory) method used to make this computation is very demanding computationally and the group is currently investigating ways of computing the change of work function on more complex geometries.

If these ideas are further validated, enhanced field emission would thus occur at locations with both geometrical enhancement and reduced work function. At the likely O(10) nm scale of field emission sites, it is likely that irregular geometry and change of work function occur together – an irregular surface inherently has a work function different than a uniform one. This insight has a practical consequence for the CLIC accelerating structures as it may explain the effectiveness of the very high-temperature treatment for the structures. The baseline fabrication procedure for structures includes a 1030°C heat treatment in a 1 bar hydrogen atmosphere, at which time the structure disks are bonded together. The hydrogen certainly makes the surface extremely clean, which is certainly important. However, the heating also fully anneals the copper. This makes it much softer, which might be expected to reduce the gradient potential of the structure. However the annealing reduces the dislocation density and consequently the number of emission sites. This may be the major reason why heating turns out to be so effective.

More weight to the idea that dislocations are the starting point of breakdowns comes from the success of a model which derives breakdown rate as a function of gradient based on the enthalpy of dislocation formation [26]. The model predicts the observed breakdown behavior very well. The two fitting parameters are dislocation formation energy and volume, and fits give physically reasonable values. The E^{30} fit described earlier in the paper work is quite accurate and convenient however does not have a physical basis. Experiments to probe a very wide range of breakdown rates to determine and eventually distinguish amongst breakdown rate dependencies using the CERN dc spark system are now underway. A new high-repetition rate pulser has been developed [28]. Preliminary data is shown in Fig. 11.

One of the main features of the evolution of a breakdown is a dramatic rise of electron current from field emission levels to many amps of current. In the CERN dc spark system this can be directly measured in the circuit and in the rf measurements through the faraday cup signals and deduced through the effect

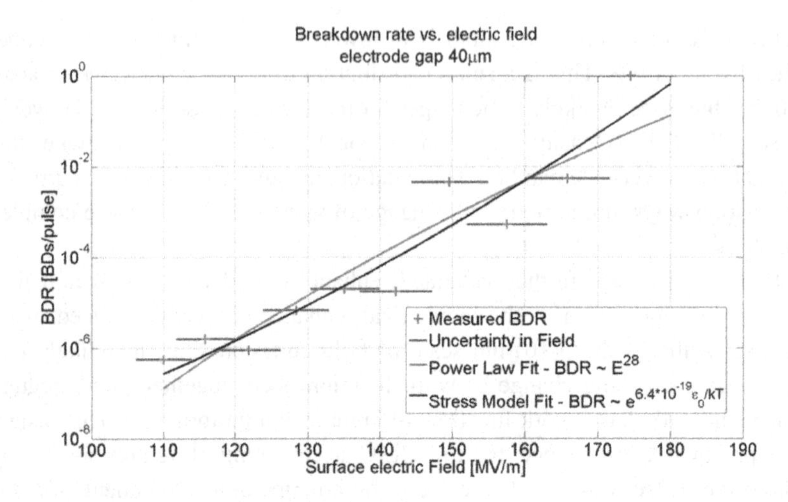

Fig. 11. Breakdown rate as a function of surface electric field in the CERN high-repetion rate dc spark experiment [29]. The exponential best fit of E28 is very similar to that seen in rf measurements. One cannot distinguish the relative validity of the two fits from this data.

on the rf signals. The incident power to the accelerating structure is quite high, approximately 50 MW in the example shown there. If currents associated with breakdown are accelerated across a single cell of the structure and then collide with the surface the approximate energy gain of an electron would be 250 MV/mx8mm = 2MeV. The current emitted from the breakdown would consequently need to exceed 20 A to absorb the incoming 50 MW. Of course the currents pull the fields lower, so the total acceleration is likely to be lower and the total current correspondingly higher.

In order to study the evolution of a breakdown from the initial atomic level trigger to times where macroscopic effects are observable, our collaboration has developed a PIC (Particle In Cell) simulation code to model the breakdown ignition process for the dc spark system geometry [30]. The code, ArcPIC, includes electric field calculation, field emission, neutral atom emission, ionization, scattering and modeling of an external driving circuit. The ion bombardment simulated in an early 1D version of the ArCPIC code was used to compute surface morphology [31].

A key simulation output is current rise as a function of time is shown in Fig. 13. Corresponding measurements of the current rise time, along with voltage fall time, have also been made for the CERN dc spark system [32]. An example is shown in Fig. 12. So far the simulations show a much faster turn on time than observed in the experiments. It is clear now that the powering circuit, with its distributed inductances and capacitances, must be included in the model in order to make a direct comparison with experiments. For example, in the 0.4 ns turn on time shown in Fig. 12, only the charge and energy stored within 12 cm can feed the arc. This energy is dominated by the gap capacitance. These energy storage effects must now be included in the simulation to compare to the measured voltage and current signals.

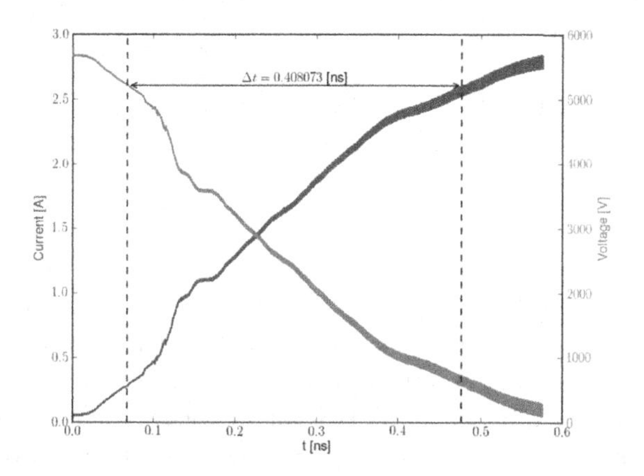

Fig. 12. ArcPIC result for a 6 μm gap [28]. As the current in the arc rises as the conducting plasma is formed, the voltage across the gap falls. A series resistance of 5000 Ω is assumed in this simulation.

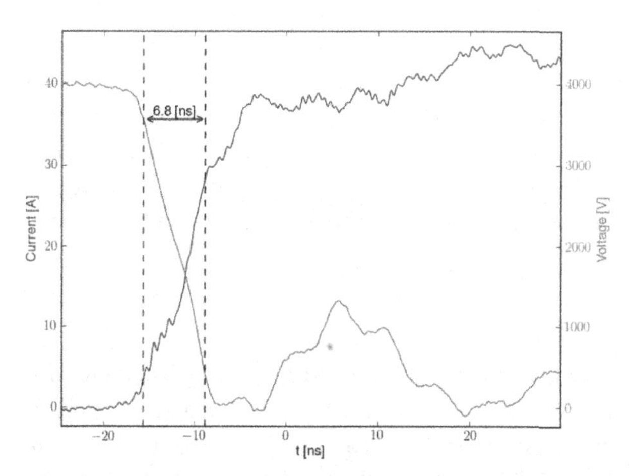

Fig. 13. Breakdown turn on time in the CERN dc spark system [27]. The current rise and voltage fall times appear to be determined by the gap and local stray capacitances of the dc system.

Table 2. Breakdown turn on times in various experiments.

Test	Frequency	Measurement	Result
Simulation dc spark	dc	Current rise time	0.4 ns
DC spark System	DC	Voltage fall and voltage rise time	7 ns
Swiss FEL (C-Band)	5.7GHz	Transmitted Power Fall Time	110–140ns
KEK T24 (X-Band)	12GHz	Transmitted Power Fall Time	20–40ns
CTF/TBTS TD24 (X-Band)	12GHz	Transmitted Power Fall Time	20–40ns
CTF SICA (S-Band)	3GHz	Transmitted Power	60–140ns

These considerations already show the importance of the powering circuit in determining the behavior of an arc. To see how the characteristics of an rf structure may affect the powering of an arc, data has been collected of the breakdown turn on time in a number of rf systems, Table 2. The origin and significance of the different times is being considered and may offer a new perspective on the high-power scaling laws. The basic idea we have now is that whether or not an arc develops may depend on the ability to feed an arc transient.

4. Summary

The effort to increase accelerating gradient in CLIC has led to significant new insights into the underlying physical processes of vacuum breakdown. These insights are now being used to direct continuing efforts to further improve structure performance and improve manufacturing techniques. The high-gradient technology is also being considered for a number of new accelerator applications.

Acknowledgments

The author first wishes to thank Juwen Wang for his pioneering role in studying breakdown in accelerating structures, his strong personal support for our CLIC X-band structure test at SLAC back in 1994 [33] and his contribution to fabricating many of the structures summarized in Fig. 1. I also wish to thank Chris Adolphsen, Sergio Calatroni, Flyura Djurabekova, Alexej Grudiev, Toshi Higo, Kai Nordlund, Igor Syratchev for their insights, intellectual stimulation and intense discussions. I would like also to thank all the co-authors listed below for the hard experimental and simulation work on which the progress is based.

References

1. M. Aicheler et al., "A Multi-TeV Linear Collider Based on CLIC Technology," CERN 2012-007.
2. A. Degiovanni, S. Doebert, W. Farabollini, A. Grudiev, J. Kovermann, E. Montessinos, G. Riddone, I. Syratchev, R. Wegner, W. Wuensch, A. Solodko B. Woolley, "High-Gradient Test Results From a CLIC Prototype Accelerating Structure: TD26CC," to be published IPAC 2014.
3. T. Higo, T. Abe, Y. Arakida, S. Matsumoto, T. Shidara, T. Takatomi, M. Yamanaka, A. Grudiev, G. Riddone, W. Wuensch, "Comparison of High-Gradient Performance in Varying Cavity Geometries," Proc. IPAC 2013.
4. C. Adolphsen, "Normal-Conducting RF Structure Test Facilities and Results", Proc. PAC 2003, Portland, Oregon, USA (2003).
5. A. Grudiev, S. Calatroni, W. Wuensch, "New local field quantity describing the high gradient limit of accelerating structures", Phys. Rev. ST- Accel. Beams 12, 102001 (2009).
6. T. Higo, private communication.
7. B. Woolley, http://indico.cern.ch/conferenceDisplay.py?confId=284650.
8. I. Wilson, W. Wuensch, C. Achard, "The fabrication of a prototype 30 GHz accelerating section for CERN linear collider studies," CERN-SL-90-84-RFL; CLIC-Note-114.- Geneva 1990.
9. J.W. Wang *et al.*, "Fabrication Technologies of the high gradient accelerator structures at 100 MV/m range", IPAC 2010, Kyoto, Japan.
10. C. Adolphsen, "Advances in Normal Conducting Accelerator Technology From the X-Band Linear Collider Program", Proceedings PAC2005.
11. W. Wuensch, "The Scaling of the Traveling-Wave RF Breakdown Limit" CERN-AB-2006-013; CLIC-Note-649, 2006.
12. M. Kildemo, "New spark-test device for material characterization" Nucl. Instrum. Methods Phys. Res., Sect. A.
13. A. Descoeudres, Y. Levinsen, S. Calatroni, M. Taborelli, W. Wuensch "Investigation of the dc vacuum breakdown mechanism" Physical Review Special Topics – Accelerators and Beams 12, 092001 (2009).
14. A. T. Perez Fontenla, private communication.
15. J.W. Wang and G.A. Loew, "Field Emission and RF Breakdown in High-Gradient Room-Temperature Linac Structure" SLAC-PUB-7684, 1997.
16. B. Jüttner, Beitr. Plasmaphys. **19**, 25 (1979).
17. G. A. Mesyats and D.I Proskurovsky, "Pulsed Electrical Discharge in Vacuum" (Springer, Berlin 1989).

18. R. Latham, "High voltage vacuum insulation: basic concepts and technological practice" Academic Press, 1995.
19. A. Descoeudres, T. Ramsvik, S. Calatroni, M. Taborelli, W. Wuensch "dc breakdown conditioning and breakdown rate of metals and metallic alloys under ultrahigh vacuum" Physical Review Special Topics – Accelerators and Beams 12, 032001 (2009).
20. A. Descoeudres, F. Djurabekova and K. Nordlund "dc Breakdown Experiments with Cobalt Electrodes" CLIC Note 875, 2009.
21. R. H. Fowler and L. Nordheim "Electron Emission in Intense Electric Fields" Proc. R. Soc. Lond. A, 119:173-181, 1928.
22. J. Frantz, M. Rusanen, K. Nordlund, and I. T. Koponen, "Evolution of Cu nanoclusters on Cu(100)", J. Phys.: Condens. Matter 16, 2995 (2004).
23. C. Kittel, "Introduction to Solid State Physics".
24. H. Chen, Y. Du, W. Gai, A. Grudiev, J. Hua, W. Huang, J. G. Power, E. E. Wisniewski, W. Wuensch, C. Tang, L. Yan, Y. You, "Surface-Emission Studies in a High-Field RF Gun based on Measurements of Field Emission and Schottky-Enabled Photoemission" Phys. Rev. Lett. 109, 204802 –14 November 2012.
25. F. Djurabekova, A. Ruzibaev, E. Holström, and M. O. Hakala," Local changes of work function near rough features on Cu surfaces operated under high external electric field". Jour. of Appl. Phys. 114, 243302 (2013).
26. K. Nordlund and F. Djurabekova, "Defect model for the dependence of breakdown rate on external electric fields" Phys. Rev. ST Accel. Beams 15, 071002 , 2012.
27. Soares, R H (CERN); Barnes, M J (CERN); Kovermann, J (CERN); Calatroni, S (CERN);
28. Wuensch, W (CERN), "A 12 kV, 1 kHz, Pulse Generator for Breakdown Studies of Samples for CLIC RF Accelerating Structures" Proc. IPAC2012.
29. N. Shipmann, private communication.
30. H. Timko, "Modelling Vacuum Arcs: From Plasma Initiation to Surface Interactions" University of Helsinki Report Series in Physics, HU-P-D188, 2011.
31. F. Djurabekova, J. Samela, H. Timko, K. Nordlund, S. Calatroni, « Crater Formation by Single Ions, Cluster Ions and Ion Showers », Nucl. Instr. And Meth. B (2011).
32. N. Shipman, S. Calatroni, R. M. Jones, W. Wuensch, "Measurement of the dynamic response of the CERN DC spark system and preliminary estimates of the breakdown turn-on time" Proc. IPAC2012.
33. J. W. Wang, G. A. Loew, R. J. Loewen, R. D. Ruth, A. E. Vlieks, I. Wilson, W. Wuensch, "SLAC/CERN high gradient tests on an X-band accelerating section" Proc. PAC95, Dallas.

X-Band Structure Development at KEK

Toshiyasu Higo

KEK/Sokendai, High Energy Accelerator Research Organization
1-1 Oho, Tsukuba, Ibaraki 305-0801, Japan

X-band accelerator structure developmentat KEK has been driven targeting the linear collidersin worldwide collaborations. It is based on the technologies developed with high-precision machining, precise assembly and bonding method to preserve the precision. With maximally utilizing the merits of such technologies, the long-range wakefield was suppressed in parallel to realize the high gradient. The latter needs more study and development to actually realize the stable operation at a gradient of 100 MV/m or higher in the view point of the present paper. The worldwide collaboration studies are extensively on-going and the understanding of the vacuum breakdown has been advancing. By describing the development at KEK toward the X-band wakefield suppressed high-gradient accelerator structure, this paper shows how such structures have been evolved and may serve to show a room for the future studies.

1. Introduction

The accelerator structure development at KEK has started from 1980's at S, C and X-band frequency range [1]. The assumed gradient level was 100 MV/m. The X-band development has extensively started in early 1990's taking the highest efficiency in the three frequencies and based on the understanding that the engineering level of high precision machining could meet the requirement of the X-band component fabrication [2].

The first X-band accelerator structure was made by CERN with using a high-precision diamond turning followed by the brazing. Due to the low group velocity of the structure with only 2% of light velocity, the beam hole aperture was 6mm in diameter. Its high gradient performance seemed satisfactory in the view point of those years, because the maximum field reached 100 MV/m [3]. It is to be noted that we did not have the firm criteria then, such as the low-enough breakdown rate.

After this initial period, the X-band development has come into the stage to make the accelerator structure with the practical length equipped with the suppression mechanism of the wakefield. The length of such accelerator structures has started from 1.3m followed by 1.8m, then 0.6m and finally

0.2~0.3m to date. The design gradient varied accordingly from 70 MV/m, then 50 and finally 100. The development has been conducted under the international collaborations with many laboratories, especially with SLAC and CERN. This history has passed the stages, firstly the high precision fabrication, then the wakefield management and finally the high gradient realization.

In the present paper are reviewed these developments of X-band accelerator structure for the linear collider at KEK. This review of the historical development reflects the evolution of the technology choices for the linear collider with considering the high-gradient and the wakefield suppression in the high-precision accelerator technology toward linear accelerator applications.

2. 0.2m Brazed Structure

2.1. *CERN made with high precision and brazed*

The structure was constant impedance design without wakefield suppression. The cells were diamond turned followed by the silver brazing. The fabrication was all performed by CERN [4] and this was the starting point of all the international collaborations related to the X-band activities at KEK.

2.2. *High gradient test at KEK*

The high gradient test was performed at Nikko experimental hall of TRISTAN Main Ring of KEK. The solenoid focus klystron with pulse-compressed by SLED was used to test the structure aiming at 100 MV/m.

It reached the target gradient level with the non-flat SLED pulse. From this result, we understood that the realization of 100 MV/m level seems fairly feasible in the X-band frequency. It was thought basically consistent to the famous Loew-Wang paper on frequency scaling on high gradient [5].

2.3. *Next development of X-band structure*

From the above understanding, we thought that the next issue after high gradient was how to suppress the wakefield, in both short-range and long-range, because the wakefield is fairly high in the X-band structure due to the small dimensions. Then, we thought that the precision fabrication technology was needed to realize the suppression. In this way, we have started seriously the high-precision machining study [6].

3. 1.3m Detuned Structure

3.1. *High precision engineering*

The industry has already reached the nano-meter-level precision machining. The technology was however on one surface, such as mirror for telescope or the semi-conductor wafer production. However, the accelerator cell, which was the sliced acceleration-cell unit, needed to be shaped over almost all of the surface including not only the inside area facing vacuum but also the outside to realize the high precision of the assembly.

We have made a dedicated ultra-precision engineering laboratory at KEK [6]. The area under the HEPA filter was temperature controlled within ± 0.5 degrees Celsius and the cleanness of the room is designed to be 10,000 particles (>0.5 μm)/cubic feet. The numerically controlled machines were sitting in the clean room, where the high volume inlet of the fresh air was required. This was because of exhausting the spent air mixed with kerosene lubricant required for the smooth cutting of copper by diamond tool.

3.2. *High precision diamond turning*

Two diamond turning machines, one with air spindle and another with more rigid spindle using static-oil pressure, were used to shape the cells. The surface finish became easily the Ra of 25 nm level. However, the dimension control was not easy because the temperature quickly changed due to the machining itself with spraying the oil mist. The outer diameter was used to control the radial dimensions, while the vacuum chucking face was refreshed to keep the reference position in the beam axis direction [6]. These two references were used to fully control all the cell dimensions, where any additional dimension accuracy was assured essentially by the numerically controlled tool movement.

In order to realize the precise dimension, it was critical to keep the vacuum chucking surface well flat. To this end, the vacuum chucking face was improved after many trials. The surface flatness was always checked by a commercially available interferometer device. Usually the flatness value better than 0.5 micron over 60-80mm in diameter was set as the reference value in most of the cases.

Milling was also done with precision milling machine with air spindle. In Fig. 1 are shown the typical accelerator cells which KEK has produced in two decades.

Fig. 1. Typical accelerator cell types which KEK has produced to date. These are chronologically aligned from top left to down right.

3.3. *High precision assembly*

The accelerator structure cells were assembled on a precise V-block. Here the precision of the outer diameter was essential. The cell-to-cell alignment was measured with two gap sensors (Microsense) running parallel to the V-block and measuring the gap distance to the outer diameter of each cell. In Fig. 2 are shown the cells stacked on the V-block for the precise alignment purpose (left) and the RF checking purpose (right). We could also measure the inclination of

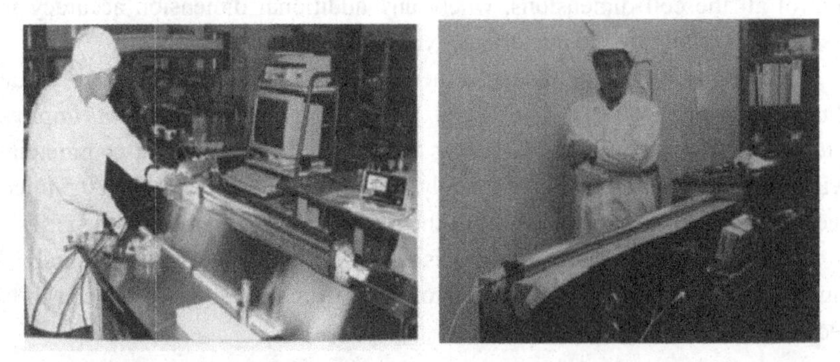

Fig. 2. Cells are precisely stacked on V-block for precise alignment by Y. Higashi (left) and for electrical checking by J. Wang (right).

cells, so-called "book-shelf", by continuously scanning the gap along the stacked cells. The bow along the whole stack was also measured by this scanning over a structure.

Once stacked well, we bonded the cells without deteriorating the cell-to-cell alignment and the bow. We do not want the cells mutually move during the bonding stage. This requirement made us develop the solid-phase bonding process for the main body of the structure. We realized a few micron level cell-to-cell misalignment and about 10 micron bow over the 1.3m structure [7]. The associated bonding technologies are described in the following section.

3.4. *High precision bonding*

To realize the solid-phase bonding technique, we have developed a so-called diffusion bonding at high temperature [8]. This development started with the idea of diffusion-brazing with a thin gold sheet between two copper cells. At about 900 degrees Celsius, some part of the contact position between gold and copper becomes fluid but quickly comes back to the solid once the gold/copper ratio changes. This process does not pass the phase with completely liquidized junction but only partly liquidized so that the slippage is suppressed. In the second step, we made the gold evaporation by 1 micron to replace the gold sheet. Finally we found that the copper-to-copper surface without any insert can still work to make the diffusion bonding at high temperature such as 800–900 degrees Celsius [9].

The basic mechanism is not well understood quantitatively. However, the parameters such as 900 degrees Celsius with 0.1MPa for 1 hour made the bonding to be well vacuum tight. However sometimes, we observed a significantly large gap, such as a few to several tens of microns, between cells, sometimes being accompaniedby the vacuum leakage, especially for the larger diameter case of 80mm.

This diffusion bonding technique was also applied at SLAC with using hydrogen furnace. At SLAC, the temperature parameter was 1040 degrees Celsius. In the standard fabrication procedure under KEK/SLAC collaboration, the SLAC parameter choice has been kept. It seems we do not have the above-described problem.

3.5. *High precision brazing*

The input and output wave guides were mounted on the stacked cells. In this case, the wave guide expands over two or three cells. Therefore, the outer diameter of the cells should be well precise and the diffusion bonding needs to keep the cell stack alignment well. By keeping the steps between cell outer surfaces within several micron level, we could mount the wave guide of the same curvature. The silver coating of 10 microns or so could realize the silver brazing as shown in Fig.3 [10].

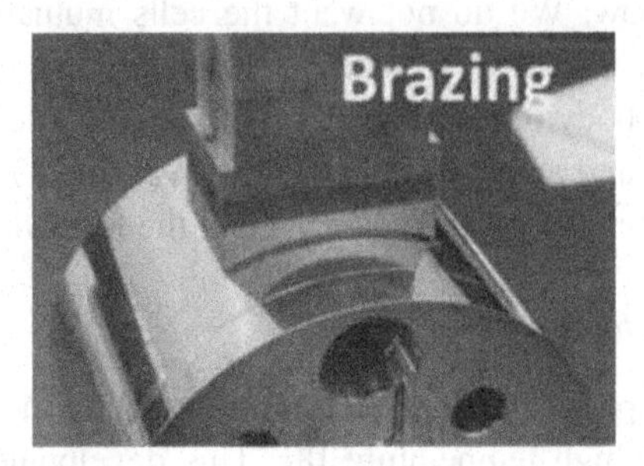

Fig. 3. Wave guide mounted on diffusion-bonded cells for brazing. A 10 micron silver is coated on wave guide side and then precision machined to fit to the outer diameter of cells.

This technique was tried in 0.3m structures and also applied to the 1.3m ones. With the latter structures, we confirmed the precision of cells and alignment by measurement of its wakefield at ASSET of SLAC [11]. One of the 1.3m structures was also tested in high power at SLAC, as described in the following section.

3.6. *Door opened to high gradient issue*

The M2 structure was high gradient tested at the klystron laboratory of SLAC. It showed the feasibility of the high accelerating field up to 85 MV/m but also showed the degradation of phase advance along the structure [12]. The change in phase advance was very large and we became aware of the serious problem associated in the high gradient operation. The issue to realize the stable high gradient has been the long-living issue as long as we are targeting the operation at the top edge of the high gradient. Actually this study has long been performed to date and is discussed in the following chapters.

4. 1.8m Damped-Detuned Structure

4.1. *Collaboration with SLAC for 1.8m structure*

We have started a formal collaboration between SLAC and KEK from middle 1990's under "International Study Group" for linear collider [13]. Since the detuned structure which we have developed in 1.3m structure was expressed as "a life sitting on a sword", the wakefield suppression method was thought dangerous from the view point of the long-term stabile wakefield control. In this respect, KEK accepted the "damped-detuned" design for the realization of wakefield suppression. This scheme keeps high in efficiency of power transfer to beam with keeping the wakefield well controlled. We also accepted the SLAC optimized length of the structure, 1.8m, where the number of couplers was reduced by taking the length of the accelerator structure as long as possible.

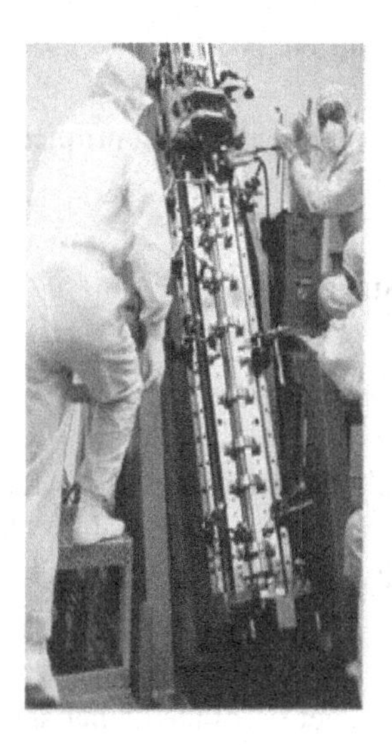

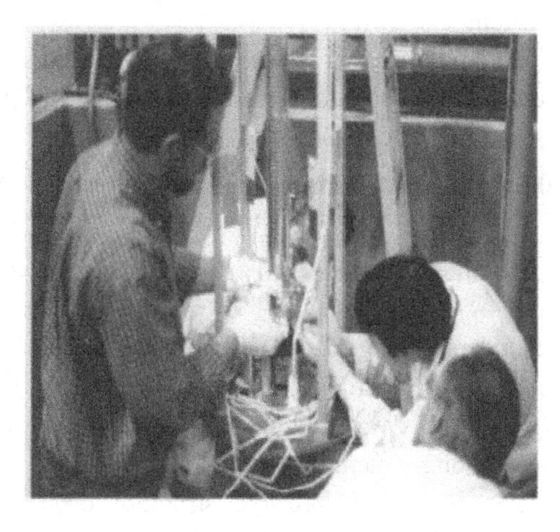

Fig. 4. Assembly of 1.8 accelerator structures. Left: Precision stacking of cells along a V-block in Japan. Right: Brazing in hydrogen furnace at SLAC.

By fully utilizing the high precision techniques developed for 1.3m detuned structures, we made a few 1.8m structures [14]. Here KEK made all the cells and made the main-body assembly followed by the diffusion bonding, while SLAC made the final brazing process to complete the structures. In Fig. 4 are shown the precision stacking stage in Japan (left) and the brazing setup stage at SLAC (right). The wakefields of these structures were measured and we confirmed the precision of the technology which met the wakefield requirement.

4.2. *Wake-field suppression in RDDS1*

One of the typical examples of the wakefield control was realized in rounded damped detuned structure (RDDS1). Again this structure was made in collaboration with SLAC and KEK, where KEK made cells and main body assembly, while SLAC made brazing and wakefield measurement.

In this structure, we made the cells without tuning of accelerating mode after assembly [15]. This is because the dimple tuning would deteriorate the higher order mode (HOM) frequency of the cells. As one of the solutions, we started the cell production from upstream side and measured accelerating mode frequency. Here we stacked 6 cells and measured the HOM frequencies at the same time to confirm the HOM behavior. Once the cumulative phase advance of the accelerating mode was observed, we made the gradual feed-forward correction to the later fabrication of cells by changing the NC value of the cell dimension.

The structure was successfully made and its wakefield was confirmed to coincide with the theoretical estimate very well [16].

5. 0.6m Damped-Detuned and High-Gradient Structure

5.1. *Problem in high gradient made us design 0.6m structure*

Once understanding of the deterioration in M2 high gradient, other 1.8m DDS structures were also confirmed to be subject to the same deterioration. We thought that this was due to the high group velocity design, which was needed for such a long structure as 1.8m long one.

Then we came into the structure design with shorter (smaller group velocity) but keeping beam hole size large in order to get rid of the higher short-range wakefield. We chose for this purpose the higher phase advance, $5\pi/6$ per cell, rather than the usual $2\pi/3$ mode. The fabrication technology was essentially the same as 1.8m structures.

5.2. *Proof of high gradient at 65 MV/m unloaded*

Here we collaborated with SLAC in a very tight collaboration framework. KEK made all of the cells but now the assembly was also done by SLAC. The processes at SLAC included the chemical etching process and the diffusion bonding in hydrogen furnace. Since these structures were mostly for high gradient evaluation, the cell stacking was assured by self-fitting feature and the frequency of accelerating mode was dimple-tuned after brazing. The final process was the vacuum baking in a vacuum can with double evacuation system for longer than 1 week at 650 degrees Celsius. Up to date, this series of fabrication processes became the standard structure fabrication method established by SLAC and KEK [17].

The high gradient test was performed mostly at NLCTA. We confirmed that the structures could be operated stably at 65 MV/m unloaded with satisfying the breakdown rate requirement of GLC [18]. This gradient is equivalent to 50 MV/m in a linear collider under a 1A current loading.

5.3. *Interleaved short structures*

Due to the limited number of cells in this 0.6m structure, or equivalently the limited number of HOM frequencies in a structure, the wakefield re-coherence time was within a bunch train. In order to get rid of this difficulty, we made the interleaving of structure HOM frequencies among 4 structures. This interleaved structure design was confirmed to suppress the wakefield well long in time by delaying the re-coherence time. Here the structure fabrication technology was the same as that of 1.8m one.

The interplay between the detailed design simulation of HOM and precision fabrication technology became one of the key technologies for the wakefield suppressed accelerator structures such as those being developed for CLIC [19].

5.4. *Structure BPM*

The long-range wakefield suppression was discussed in the above sections. We think that the technology to realize has been well established.

In turn, it is essential to align the structures to the beam to reduce the short-range wakefield and the requirement is of the order of several microns. It is difficult to do it with the alignment technique using the information outside the structure. The damped-detuned structure design makes it possible for us to detect the beam position by using the beam-excited signal which is leaked into the manifold running along the structure. By carefully selecting the frequency component of the signal, which is related to the position of the cells for the HOM of the relevant frequency, we can deduce the position information of the beam at those cells. This technique, structure BPM, was proven in 1.8m structures and 0.6m structures [20].

6. Toward Higher Gradient for CLIC Main Acceleration

6.1. *Shifting target toward higher gradient*

The ITRP (International Technology Recommendation Panel) in 2004 has chosen the superconducting RF technology for linear collider [21]. Form this decision, the X-band developments at KEK became independent of the linear collider, but exploring more to the application of lower energy machines. However in late 2006, the CLIC (Compact Linear Collider) which CERN was developing at 30 GHz has re-optimized its frequency down to X-band (12 GHz) with the gradient at 100 MV/m for reaching the c.m. energy of 3 TeV [22].

Since the main linac acceleration technology was very close to that we have been developing, we jumped into the collaboration with CLIC. The strategy was to fully utilize the legacy of the technology background developed with SLAC. One of the big differences from GLC was the high acceleration gradient of 100 MV/m. Though the shorter accelerating structure with smaller group velocity was supposed to be preferable for the stable higher gradient, we were not sure whether such high gradient could actually be realized. The required breakdown rate (BDR) for CLIC was 3×10^{-7} breakdown per pulse per meter (bpp/m) and it requires a month or more to prove the feasibility in the statistically meaningful manner. This requirement is equivalent to less than 1 breakdown per 4 days in 0.2m accelerator structure operated at 50 Hz. The long-term running test is essentially important so that the Nextef of KEK was established to pursuit the program [23].

6.2. *Structure fabrication and high gradient test*

Since the standard fabrication technology of SLAC/KEK was thought to be the best method to start the study, we decided to follow exactly the same framework before ITRP, i.e. KEK made all the relevant accelerator components for a pair of structures, while SLAC made the twin through the cell cleaning, diffusion bonding, brazing, tuning and finally the vacuum baking. Example structures are shown in Fig. 5, where the different wave guide flanges indicate the demand from two laboratories.

The high gradient test was programmed at both laboratories using one out of the pair of structures. The study became much reliable by pursuing the evaluation with comparing the two independent tests on those made in exactly the same fabrication technology and at very close in time [24]. Results to date are described below, but we report here mostly those from Nextef of KEK.

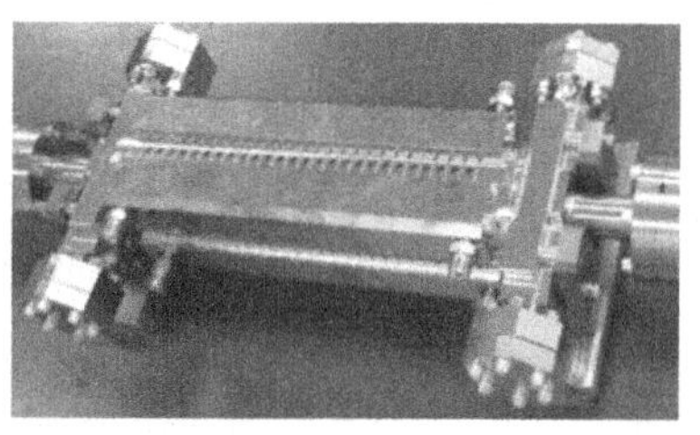

Fig 5. Twin structures designed, made and tested by CERN, SLAC and KEK. Left figure for test at SLAC and right onethat at KEK.

6.3. *Typical features of the high gradient operation*

The processing is usually performed starting from the short pulse operation, typically 50 nsec. We raise the peak power for the processing with monitoring the vacuum pressure level in addition to checking the pulse shape change. Once the pressure rises or RF shape changes, the operation stops and we wait about half a minute before restarting at a somewhat lower power level [23].

This processing needed time, especially the first and shortest pulse length. Those at the later time and with the longer pulse lengths became much faster to be conditioned till finally reaching to 240 nsec, the goal for our case which was determined as the total pulse length for CLIC operation including the ramping time. During the initial processing stage, the dark current became less and less. After reaching the goal parameters, the breakdown rate (BDR) went down as the processing proceeds. In the following sections, we review some cases showing the typical features relevant to the high gradient performance.

One of the important features was found and analysed with the idea of complex pointing vector [25]. In this idea, the field emission electrons under electromagnetic field either in standing wave or in travelling wave can be taken into account. The recent structure designs for CLIC travelling wave were based on this criterion. Therefore, the test structures which KEK has been studying these years in collaboration with CLIC are basically based on this.

6.4. *High-performance example*

The best performance was observed in "T24#3"as shown in Fig.6[26]. It is the second series of CLIC prototype structures, though without damping feature, with 24 accelerating cells to flatten the field profile than the first design series with 18 cells. It showed the very low BDR scaled at 100 MV/m, which met the CLIC requirement. It showed ever decreasing characteristics in BDR until the experiment was interrupted by the big earthquake in March 2011.

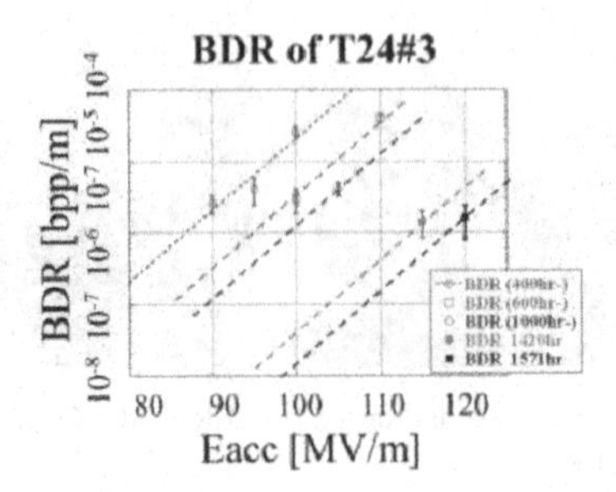

Fig. 6. Breakdown rate of T24#3 versus Eacc at various stages of the processing. The dashed slopes are just for guiding eye with taking the same slope from the earliest data with three points.

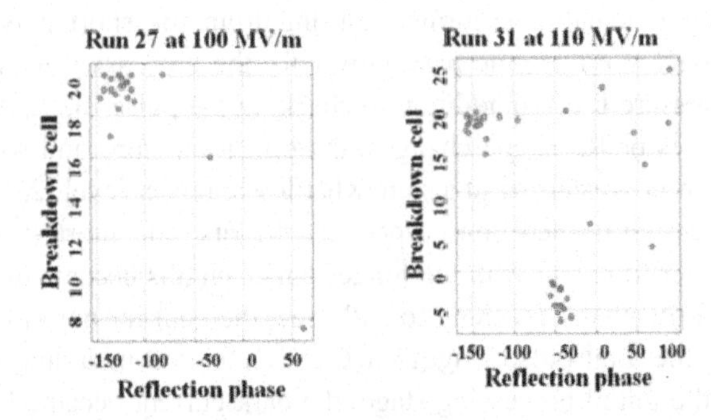

Fig.7. The cell number of the reflection phase of the breakdowns in TD24R05#2 structure test. The left figure shows the breakdown points after reaching 100 MV/m, while the right figure those after gradient increase to 110 MV/m.

6.5. *Poor-performance example*

After the experimental setup was recovered from the earthquake damage, the damped version of the second series, TD24#4 and TD24R05#2 were tested. These showed the poorer BDR possibly due to the high magnetic field [27]. It was accompanied even by the frequent breakdowns in a localized region of the structure as shown in Fig. 7 [28]. This localization of breakdown position is called "hot spot" and the BDR in this case has not been reducing but saturated, indicating that the appearance of hot spot suppressed the BDR improvement.

6.6. *Proven high gradient in waveguide damped structure*

From the experimental results on CLIC prototype accelerator structures with wave-guide damping feature, we conclude that the gradient level of 80 MV/m or so is quite feasible with the preparation technology to date. It may be applied to a medium energy linear collider such as Higgs Factory driven by conventional RF sources such as klystrons in several kilometer size [29].

We should admit that the realization of 100 MV/m or more has not been well confirmed, especially in the wakefield suppressed ones. The establishment to this end still needs some technology development for the preparation of the structure or possibly new geometries such as the choke-mode design.

6.7. *Study on quadrant*

As one of the very different design of the accelerator structure, the idea of quadrant was proposed and some tests were done at CERN. KEK also tried to taste the idea. Firstly a high-gradient test structure was made with ball-point milling with carbide tool [30].

High gradient test was performed at KEK. It showed the similar, poor, result as the test previously done at SLAC. The reason is not understood but we think it worthwhile to study it, because the engineers think this type is suited for the low-cost high-volume production.

6.8. *Perspective toward high gradient*

The gradient of 100 MV/m or higher is needed for the main linac of the multi-TeV scale high energy linear collider [22]. To this end, we definitely need to understand the vacuum breakdown mechanism to make the accelerator stable to meet the high luminosity demand.

It is necessary to understand the vacuum breakdown mechanism happening in the present prototype structures. This has been pursuing in recent years worldwide from theoretical simulation based on the first principle of quantum atomic mechanism into plasma evolution to reproduce various experimental studies. KEK has been providing the experimental studies in a series of long-term and high gradient in prototype structures. We have also prepared another setup dedicated for the basic experimental study.

The choke mode design might be one of the possible candidates for such high gradient, because it can operate without magnetic field enhancement, which is thought as the possible origin of the breakdown observed from the recent year experiments. The design for CLIC purpose has been done [31] and the prototype test structure is under development at Tsinghua University to be high-power tested at Nextef of KEK.

Basic study to understand the vacuum breakdowns is definitely needed to explore the normal conducting high gradient linac. It is needed not only for the linac requiring the very high gradient to reach very high energy (TeV scale) but also medium (GeV scale) or low energy (MeV scale) machine in case the operational stability is required. There are many pioneering studies to this direction. In the next chapter are shown some of such activities, where there are KEK contributions in various points.

7. Effort to Understand the Mechanism of Breakdowns

The collaboration with SLAC on materials has been pursued for many years. It showed hard material to suppress the RF-current-driven fatigue on copper surface [32]. The hard copper may play an essential role as far as the initial stage to be favorable to the high gradient [33].

The DC breakdown studies have been pursued at CERN. The comparison among various metals was systematically done to prioritize on the breakdown threshold [34]. In recent years, the more sophisticated experimental observations were tried, such as measuring the field-emission current to observe pre- or post-breakdown phenomena.

At KEK, another experimental setup, Nextef shield-B, was prepared in addition to the present Shield-A, where the CLIC prototype structure tests are usually been performed [23]. Here we are preparing such test as the study on the mechanism of poor high gradient observed in quadrant type. We hope here to address such studies as those addressing thematerial choice, preparation of surface and the base material under surface of interest.

8. Conclusion

KEK has been developing the X-band accelerator structures for linear collider of TeV energy range. Using the high-precision engineering with diamond turning, diffusion bonding and so on, we confirmed the feasibility of wakefield suppression. The gradient of 80 MV/m level seems quite feasible but that at 100 MV/m or higher seems to requires more basic understanding of vacuum breakdown mechanism and the suppression technology.

The collaboration among CERN, SLAC and KEK has been playing an essential role in this development. The study toward the gradient at 100 MV/m or higher still needs the basic physical understanding of vacuum breakdown mechanism. We want to keep studying in this trend in addition to explore the possibility of applying the technology which we have been developed during two decades.

Acknowledgments

Firstly we greatly thank SLAC for developing the X-band technology with KEK in the first decade of the development till early 2000. In this period the highest acknowledgment should be paid to Dr. Juwen Wang of SLAC who initiated the intimate collaboration in structure development and always managed the important processes at SLAC to realize the critical pass in the fabrication. We are much indebted to Dr. C. Pearson and his colleagues, who taught us the fabrication practice and conducted the excellent fabrication works. We also enjoyed the structure meeting at SLAC lead by Juwen and it was the source of new ideas for the critical issues. Around the meeting, we enjoyed the stimulus discussions and learned many from Professors G. Loew, R. Miller, P. Wilson, N. Kroll and many others with the deep knowledge based on SLAC two mile Monster. The financial support here was basically under US-Japan cooperation program.

In turn, we thank CERN for providing us with a structure made by the high precision technology, from the beginning of KEK structure development, actually in 1992. Especially Dr. Ian Wilson was the keyto establish the collaboration, which has been continuing to date. Since 2007, this collaboration activities have been under "Agreement of Collaboration Work", ICA-JP-0103, as CLIC collaboration between CERN and KEK.

We want to express sincere thanks to the colleagues of Protvino branch, Budker Institute of Nuclear Study, Russia, where 14 GHz activities were ongoing in early 1990's. We learned not only their high precision technologies applied to the mass production of accelerator structures but also many study activities till today through those such as Drs. V. Dolgashev of SLAC, I. Syratchev of CERN and S. Kazakov of FNAL.

Finally and not least, we thank KEK management to promote the study before ITRP and kept supporting to date. Sincere thanks should be given to Dr. Y. Higashi and T. Takatomi of Mechanical Engineering Center of KEK, who have been leading the high-precision technology for accelerator structure development through the whole history of the X-band development at KEK.

All these people contributed to the successful achievement of the present wakefield-suppressed high-gradient accelerator structure for linear collider. I believe such collaboration is still beneficial and necessary for the actual advancement of the X-band technology toward future advance accelerators. I feel such collaboration as that between Tsinghua University and KEK, established in 1990's by Professors Y. Lin of Tsinghua University and K. Takata of KEK, has been playing one of the key collaborative works in Asia. Here I find again the contribution of Dr. Juwen Wang who promoted the Tsinghua's, Chinese and worldwide effort. We greatly thank him for all the activities over more than two decades

References

1. JLC-I, *KEK Report 92-16*, December, 1992.
2. JLC Design Study, *KEK Report 97-1*, April, 1997.
3. T. Higo et al., "High Gradient Performance of X-band Accelerating Sections for Linear Colliders", *Part. Accel.*, **48**, pp.43–59, 1994.
4. T. Higo et al., "High-gradient Experiment on X-band Disk-Loaded Structure", *KEK Report 93-9, CERN SL/93-39*, September, 1993.
5. G. Loew and J. Wang, "RF Breakdown Studies in Room Temperature Electron Structures", *SLAC-PUB-4647,* May 1988.
6. Y. Higashi et al., "Studies on High-precision Machining of Accelerator Disks of X-band Structure for a Linear Collider", KEK Report 2000-1, May 2000.
7. T. Higo et al., "Precise Fabrication of 1.3m-long X-band Accelerating Structure", THP47, *LINAC96*, Geneva, Switzerland, 1996.
8. Y. Fukaya et al., "Diffusion Bonding of Copper Plates made by Ultra-Precision machining", *Contributed Papers of Welding Society of Japan,* in Japanese, Vol. 15, No. 3, 1997.
9. Y. Higashi et al., "Study on High-precision Diffusion Bonding for X-band Accelerator Structure", *KEK Report 2000-2*, April 2000.
10. A. Yamamoto et al., "Fabrication of an X-band 30 cm Accelerating Structure by Diffusion Brazing", *EPAC94*, London, July, 1994.
11. C. Adolphsenet al., "Measurement of Wakefield Suppression in a Detuned X-Band Accelerator Structure", *Phys. Review Letters*, **27**, p2475, 1995.
12. R. Loewen *et al.*, "SLAC High Gradient Testing of a KEK X-Band Accelerator Structure", *PAC99*, New York, USA, 1999.
13. International Study Group Progress Report, *KEK Report 2000-07, SLAC R-559*, April, 2000.
14. J. Wang et al., "Accelerator Structure R&D for Linear Colliders", FRA18, *PAC99*, New York, USA, 1999.
15. T. Higo et al., "Meeting Tight Frequency Requirement of Rounded Damped Detuned Structure", *LINAC2000*, Monterey, USA, Aug., 2000.
16. R. M. Jones et al., "The Transverse Long-Range Wakefield in RDDS1 for the JLC/NLC X-band Linacs", p3468, *PAC99*, New York, USA, March, 1999.
17. J. Wang and T. Higo, "Accelerator Structure Development for NLC/GLC", *ICFA Beam Dynamics News Letter* No. **32**, pp. 27–46, December 2003.
18. C. Adolphsen et al., "Advances in Normal Conducting Accelerator Technology from the X-Band Linear Collider Program", *PAC05*, Knoxville, Tennessee, USA, 2005.

19. R. M. Jones et al., "Influence of fabrication errors on wake function suppression in NC X-band accelerating structures for linear colliders", *New Journal of Physics*, **11**, 033013, Mar. 2009.
20. S. Doebert et al., "Beam Position Monitoring Using the HOM-Signal from a Damped and Detuned Accelerating Structure", *PAC05*, May 2005, USA, 2005.
21. ITRP, International Technology Recommendation Panel Report, http://www.esgard.org/documents/doc/ITRP_Report_Final.pdf
22. CLIC CDR, http://clic-study.org/accelerator/CLIC-ConceptDesignRep.php
23. S. Matsumoto et al., "High gradient test at Nextef and high-power long-term operation of devices", *Nuclear Instruments and Methods in Physics Research* **A657**, pp. 160–167, 2011.
24. T. Higo et al., "Advances in X-band TW Accelerator Structure Operating in the 100 MV/m Regime", THPEA013, *IPAC10*, Kyoto, May 2010.
25. A. Grudiev et al., "New local field quantity describing the high gradient limit of accelerating structures", *Phys. Rev. ST-AB*, **12**, 102001, 2009.
26. T. Higo, Presented at the Linear Collider Workshop, LCWS12, Arlington, USA, Oct. 2012, http://www.uta.edu/physics/lcws12/pages/registration.html.
27. F. Wang et al., "Performance limiting effects in X-band accelerators", *Phys. Rev. ST-AB*, **14**, 010401, 2011.
28. T. Higo, Presented at the CLIC workshop, CLIC 2013, CERN, Feb. 2013, http://indico.cern.ch/conferenceDisplay.py?confId=204269.
29. R. Belusevic and T. Higo, "A CLIC Prototype Higgs Factory", KEK Preprint 2012–21, August 2012.
30. T. Higo et al., "Fabrication of a Quadrant-Type Accelerator Structure for CLIC", WEPP084, *EPAC08*, Genoa, Italy, 2008.
31. H. Zha et al., "Choke-mode damped structure design for the Compact Linear Collider main linac", *Phys. Rev. ST-AB*, V15, 122003, 2012.
32. L. Laurent et al., "Experimental Study of RF Pulsed Heating", *Phys. Rev. ST-AB*, **14**, 041001, 2011.
33. Valery Dolgashev, "Progress on high-gradient structures", AIP Conference Proceedings 1507, 76 (2012); doi: 10.1063/1.4773679.
34. A. Descoeudres et al., "Dc breakdown conditioning and breakdown rate of metals and metallic alloys under ultrahigh vacuum", *Phys. Rev. ST-AB*, **12**, 032001, 2009.

A Compact Wakefield Measurement Facility

J. G. Power and W. Gai

ANL, Argonne, IL 60439, USA

The conceptual design of a compact, photoinjector-based, facility for high precision measurements of wakefields is presented. This work is motivated by the need for a thorough understanding of beam induced wakefield effects for any future linear collider. We propose to use a high brightness photoinjector to generate (approximately) a 2 nC, 2 mm-mrad drive beam at 20 MeV to excite wakefields and a second photoinjector to generate a 5 MeV, variably delayed, trailing witness beam to probe both the longitudinal and transverse wakefields in the structure under test. Initial estimates show that we can detect a minimum measurable dipole transverse wake function of 0.1 V/pC/m/mm and a minimum measurable monopole longitudinal wake function of 2.5 V/pC/m. Simulations results for the high brightness photoinjector, calculations of the facility's wakefield measurement resolution, and the facility layout are presented.

1. Introduction

A fundamental concern for the NLC is the beam-induced wakefield effect. In general, the short-range wake acts back on the beam and degrades its quality, while the long-range wake deleteriously affects subsequent bunches. Wakefield effects, such as those due to RMS structure misalignment and schemes to suppress these effects (e.g. emittance bumps), have been studied mainly with numerical calculations and with previous machines, such as the SLC. These techniques, however, will work at the NLC only if the structure wakefields are below certain thresholds. For example, the RMS structure misalignment must be less than 20 μm to avoid resonant instabilities. For quality control reasons, it would be beneficial if the wakefield characteristics of the NLC structures were accurately measured before being installed into NLC accelerating modules. We propose to characterize these structures by directly mapping their wakefields [1].

Direct mapping of the wake function of NLC structures was previously made at the AATF [2], with moderate precision using approximately 10 MeV beams, and also at ASSET [3], with high precision using approximately 1 GeV beams. In this paper, we present a zeroth order design study that shows that direct-wakefield measurements can be made with high precision using 10 MeV beams if the facility is based on high-brightness RF photocathode guns [4]. Such a facility could then be used as the basis of an NLC quality control center.

2. Facility Overview

A compact, collinear, direct-wakefield measurement facility [Fig. 1] can be made using a high-brightness, 20 MeV, 2 nC drive beam to excite the wakefields and a high-brightness, 5 MeV, 0.1 nC witness beam to probe the wakefields. For wakefield measurements, a beamline is designed to transport the drive and witness beams collinearly through the structure. The longitudinal location of the witness beam with respect to the drive beam can be continuously varied from a time delay t of one nanosecond ahead of the drive to tens of nanoseconds behind it. This is accomplished by varying the optical delay path of the laser pulse generating the witness beam in conjunction with an adjustment to the phase of the RF into the witness gun.

The drive beam is matched from the output of the drive gun and linac (not shown) to the drive beam lattice (quad (Q) triplet and dipole (D) chicane) and through the NLC structure. In a similar fashion, the witness beam is delivered through a dog-leg section and into the NLC structure. Initial alignment of the drive (witness) beam through the structure is done while blocking the witness (drive) beam and centering the drive (witness) beam on the straight-through zeroing BPM.

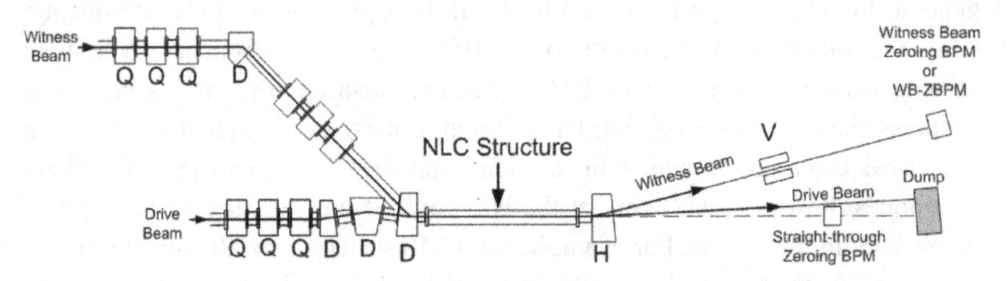

Fig. 1. Block diagram of the compact wakefield measurement facility.

2.1. *Measurement of the longitudinal wake*

Upon exiting the structure, the drive and witness beams are separated by a horizontal dipole magnet (*H*) that bends the beams by approximately 2.5° and 10° respectively. With the witness beam running in front of the drive beam, the horizontal dipole magnet, *H*, is adjusted until the witness beam is centered in the horizontal direction on the witness-beam zeroing BPM (WB-ZBPM). To measure the monopole longitudinal wake function the structure is centered (zero offset with respect to the beams) and the witness beam is positioned at a time t behind the drive beam. The longitudinal wakefield is proportional to the longitudinal momentum kick received by the witness beam,

$$\Delta\theta_z(t) = \frac{\gamma}{\gamma+1}\left(\frac{\Delta E_z(t)}{E}\right) \tag{1}$$

where γ is the relativistic energy factor, E is the energy of the witness beam (5 MeV), and ΔE_z is the longitudinal energy change of the witness beam centroid. After exiting *H*, the drive beam is dumped and the witness beam is directed towards a weak vertical dipole magnet (*V*) and the WB-ZBPM. Conceptually, one could simply measure the horizontal offset at WB-ZBPM to infer the energy change (ΔE_z) of the witness beam centroid. However, from the point of view of beam optics and BPM resolution, it is better to use *H* to center the witness beam centroid on *WB-ZBPM* and then use this change to the magnetic field of *H* to calculate the change in energy, ΔE_z.

2.2. *Measurement of the transverse wake*

To measure the dipole transverse wake function the structure is displaced 1 mm in the vertical direction relative to the collinear drive and witness beams. The witness beam is then placed at time t behind the drive bunch, which causes the trailing witness beam to receive a vertical deflection,

$$\Delta\theta_y = \frac{\gamma}{\gamma+1}\left(\frac{\Delta E_y}{E}\right) \tag{2}$$

where ΔE_y is the transverse energy change of the witness beam centroid. To compensate for the longitudinal wake, the magnet *H* is again used to horizontally center the witness beam on WB-ZBPM. The vertical deflection of the witness beam ($\Delta\theta_y$) will cause the beam to be displaced in the vertical direction (out of the plane in Fig. 1) at WB-ZBPM. Once again, conceptually, we could measure this vertical offset, at WB-ZBPM, to infer the strength of the transverse wake, but for practical purposes it is better to use the vertical dipole magnet *V*, to center the witness beam on WB-ZBPM and use the change in magnetic field strength of *V* to infer $\Delta\theta_y$.

2.3. *Machine functions*

The machine functions for matching the drive and witness beams through the structure and into the measurement area were found with COMFORT [5]. The results are shown in Figs. 2 and 3.

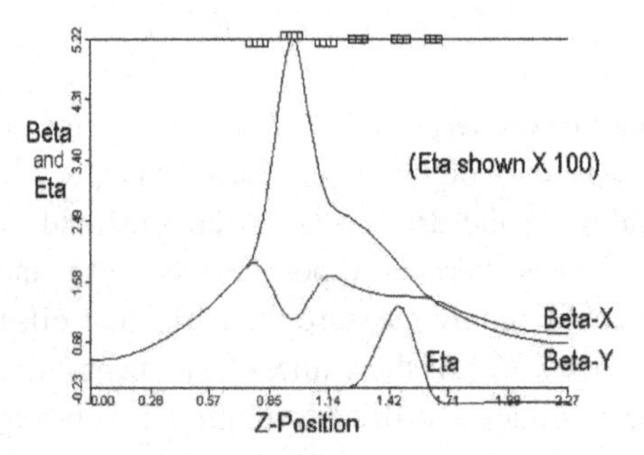

Fig. 2. Drive beam machine functions from the end of the drive linac to the center of the NLC structure.

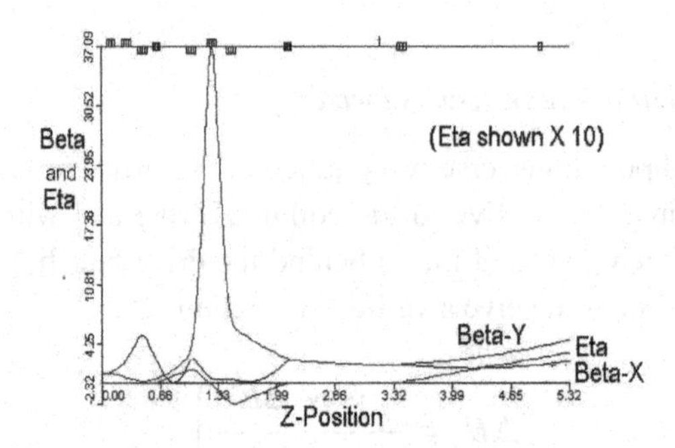

Fig. 3. Witness beam machine functions from the end of the witness gun to WB-ZBPM.

3. High Brightness Electron Source

A new 1 ½ Cell L-band (1.3 GHz) RF photocathode gun at the Argonne Wakefield Accelerator (AWA) facility has recently been commissioned. The primary purpose of this gun is to generate high-intensity beams, with a nominal bunch charge of 100 nC, for studying wakefield acceleration schemes. The beam produced by the gun in this high-intensity mode is not suitable for characterization of NLC structures due to the large normalized beam emittance of approximately 100 mm-mrad.

In order to determine if this gun can be operated in a high-brightness mode to generate the drive beam, we first spend a moment to discuss the requirements of the drive beam. The maximum allowable emittance of the electron source is estimated by examining the dimensions of an NLC structure and then requiring that the beam be able to pass through it. Typical NLC structures currently under consideration are about 1 meter long with an inner radius of the iris near 3 mm. We choose a comfortable safety margin by requiring that the beam's one-sigma radius be about an order of magnitude less than the iris radius, or $\sigma_{x,y} \approx 300$ μm. Assuming that: (1) the beam is focused to a waist at the center of the structure; (2) the beta function is one meter, or $\beta = 1$ m; and (3) the normalized energy of the beam is $\gamma = 40$; then the maximum allowable normalized emittance from $\sqrt{\varepsilon\beta} = 300\mu m$ is $\varepsilon_n^{max} = 3.6$ mm mrad.

Table 1. High-Brightness Operating Mode Parameters.

Parameter	Value
Charge (nC)	2
Laser spot radius (mm)	1.6
Laser pulse length	8
RF launch phase (^{0})	40
Energy (MeV)	10.4
Energy spread (%)	0.5%
Normalized r.m.s. emittance (mm mrad)	1.9

Our second requirement is to operate the drive gun at the highest possible charge to increase our wake function measurement sensitivity. Recent Parmela [6] simulations indicate that the gun can indeed be operated in a *high-brightness mode*. The beam parameters produced by the AWA gun operating in high-brightness mode are summarized in Table 1. It is worth pointing out that this operating point uses a drive beam charge of 2 nC that meets the drive beam requirements easily.

4. Resolution of The Wakefield Measurement System

In this section we put all the previous sections together and make an estimate of our measurement resolution.

4.1. *Longitudinal wakefield resolution*

The longitudinal momentum kick (see Eq. (1)) is related to the longitudinal wake function by,

$$\Delta\theta_z(t) = \frac{\gamma}{\gamma+1}\left(\frac{-eQ_d L_s W_{\parallel,0}(t)}{E}\right) \tag{3}$$

where Q_d is the drive bunch charge in pC, L_s is the structure length in meters, E is the witness beam energy, and $W_{\parallel,0}(t)$ is the monopole (m=0) longitudinal wake function [7] per unit length in units of *V/pC/m*. From this equation we can see that we will achieve good sensitivity to the wake function (i.e. a large kick) if Q_d is large and E is small. Since Q_d and E are fixed at 2 nC and 5 MeV, respectively, then the resolution depends on the accuracy to which we can measure the energy change, $\Delta E_z/E$.

If the witness beam charge is 10 pC, we can estimate a normalized r.m.s. emittance of 0.1 mm mrad. From Fig. 3, we see that the horizontal beta function of the witness beam is $\beta_x = 2.6$ m and the dispersion is $\eta_x = 0.2$ m at WB-ZBPM. The width due to the beta function ($\sigma_\beta = 0.5$ mm) is added in quadrature with the width due to dispersion function and 1% energy spread ($\sigma_\eta = 2.0$ mm) to give $\sigma_{tot} = 2.1$ mm. We now assume that the accuracy of WB-ZBPM is about 1/10th of the one sigma beam width, or 210 μm. Since H bends the witness beam by 10°, then a 0.1% change in longitudinal momentum over a 2 m drift will produce a horizontal offset of the witness beam centroid at WB-ZBPM of 350 μm, well within the resolution of WB-ZBPM. Finally, combining Eq. (1) and Eq. (3) and solving for $W_{\parallel,0}(t)$ gives a minimum measurable monopole longitudinal wake function of 2.5 V/pC/m.

4.2. *Transverse wakefield resolution*

The transverse momentum kick (see Eq. (1)) is related to the transverse wake function by,

$$\Delta\theta_y(t) = \frac{\gamma}{\gamma+1}\left(\frac{-eQ_d L_s W_{\perp,1}(t)}{E}\right)\Delta y_d \tag{4}$$

Where $W_{\perp,1}(t)$ is the dipole (m=1) transverse wake function per unit length in units of V/pC/m/mm, Δy_d is the offset of the drive beam relative to the center of the structure measured in mm, and all other variables have been previously defined. Once again, good sensitivity to the transverse wake function is obtained when Q_d is large and E is small.

From Fig. 3, we see that the vertical beta function of the witness beam is $\beta_y = 4.6$ m leading to a vertical beam size of $\sigma_\beta = 0.4$ mm at WB-ZBPM. Once again, we assume that our BPM resolution is 1/10th of the spot size, so that the vertical BPM resolution is 40 μm. Next, we estimate the minimum angular kick ($\Delta\theta_y$) that our system can detect. Since the drift length is 2 m, a 20 μrad kick produces an offset of 40 μm at WB-ZBPM, an amount within our resolution.

Finally, solving Eq. (4) for $W_{\perp,1}(t)$ gives a minimum measurable dipole transverse wake function of 0.1 V/pC/m/mm.

5. Conclusion

We have presented a conceptual, zeroth order design for a compact wakefield measurement facility. This facility can measure both the longitudinal and transverse wake functions with state of the art precision. Our estimates show that we can measure a minimum measurable monopole longitudinal wake function of 2.5 V/pC/m and a minimum measurable dipole transverse wake function of 0.1 V/pC/m/mm. The Argonne Wakefield Accelerator facility at ANL could be used to build a prototype version of this facility to prove the validity of this concept.

References

1. H. Figueroa et al., Phys. Rev. Lett., **60**, No. 21 (1988).
2. J. W. Wang, et al., Proc. PAC 1991, pp. 3219–3221.
3. C. Adolphsen, SLAC-PUB-6629.
4. M. E. Conde et al., Phys. Rev. ST Accel. Beams **1**, 041302 (1998).
5. M.D. Woodley, et al., IEEE Transactions on Nuclear Science, **NS-30**, No. 4, (1983).
6. L.M. Young, Physics of Particle Accelerators, **AIP-184**, p. 1245, New York, (1989).
7. A. W. Chao, “Physics of Collective Beam Instabilities in High Energy Accelerators,” John Wiley & Sons, New York, 1993.

Simulation Code Development and Its Applications*

Zenghai Li
SLAC, Menlo Park, CA 94025, USA

Under the support of the U.S. DOE SciDAC program, SLAC has been developing a suite of 3D parallel finite-element codes aimed at high-accuracy, high-fidelity electromagnetic and beam physics simulations for the design and optimization of next-generation particle accelerators. Running on the latest supercomputers, these codes have made great strides in advancing the state of the art in applied math and computer science at the petascale that enable the integrated modeling of electromagnetics, self-consistent Particle-In-Cell (PIC) particle dynamics as well as thermal, mechanical, and multi-physics effects. This paper will present the latest development and application of ACE3P to a wide range of accelerator projects.

1. Introduction

SLAC pioneering work in high-performance computing (HPC) for accelerator applications under DOE support started over a decade ago with the award of the Accelerator Grand Challenge, later became the Scientific Discovery through Advanced Computation (SciDAC) [1] program. The main goal of the research is to develop and apply high-performance computational tools for the design, optimization and analysis of existing and future accelerators. With a strong collaboration with the SciDAC Centers for Enabling Technologies (CETs) and Institutes to develop and deploy advanced applied mathematics and computer science techniques, SLAC has developed the electromagnetic simulation suite ACE3P (*A*dvanced *C*ompuational *E*lectromagnetics *3P*) [2], consisting of RF modules in frequency and time domains as well as thermal and mechanical solvers. These massively parallel codes are based on the higher-order finite-element method so that geometries of complex structures can be represented with high fidelity through conformal grids, and high solution accuracies can be

*Work supported by the U.S. DOE ASCR, BES, and HEP Divisions under contract No. DE-AC02-76SF00515. The work used the resources of NCCS at ORNL which is supported by the Office of Science of the U.S. DOE under Contract No. DE-AC05-00OR22725, and the resources of NERSC at LBNL which is supported by the Office of Science of the U.S. DOE under Contract No. DE-AC03-76SF00098.

obtained using higher-order basis functions.These tools aim to achieve virtual prototyping of accelerator structures through HPC and have had significant impact on the design, optimization and analysis of accelerator projects.

2. Parallel Higher-Order Finite Element Method

The ACE3P codes are developed based on the parallel higher-order FE method. A key advantage of the finite element conformal (tetrahedral) mesh over the finite difference structured mesh is geometry fidelity as shown in Fig. 1 of a finite element mesh for a cavity with a coaxial coupler. Further accuracy can be obtained through the use of quadratic surface and higher-order elements (p = 1-6) resulting in reduced computational cost as seen in Fig. 1. An equaling important advantage is the power of parallel processing which both increases memory and speed, allowing large problems to be solved in far less time through scalability. The success of large-scale simulations through high performance computing (HPC) relies heavily on the computational science research funded by SciDAC to improve code scalability and produce the desired results.

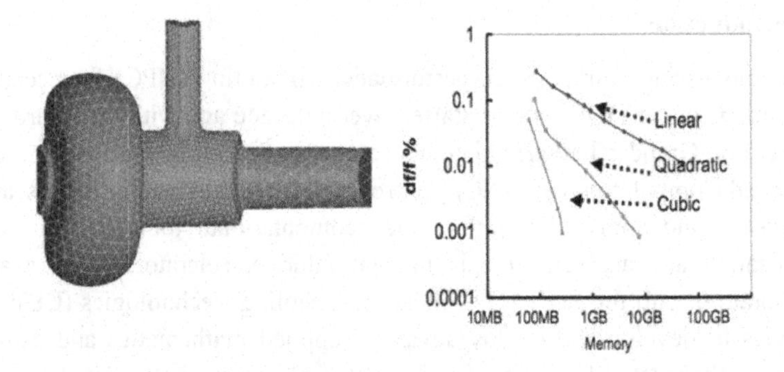

Fig. 1. (Left) Tetrahedral mesh of a coupler cavity; (Right) Frequency convergence of a cavity mode in terms of computer memory usage using p = 1, 2 and 3 basis functions in elements.

A suite of parallel codes in production mode has been developed and includes:

Omega3P: a complex eigensolver for finding normal modes in resonant structures with open ports, impedance boundaries or lossy materials;

S3P: a frequency domain solver for calculating scattering parameters of RF components;

T3P: a time-domain solver for simulating transient response of RF driven systems and for calculating wakefields due to charged beams;

Track3P: a particle tracking code for calculating dark current and analyzing multipacting in RF cavities and components;

Pic3P: a particle-in-cell code for self-consistent simulation of particle and RF field interactions in RF guns and klystrons;

TEM3P: an integrated multi-physics code including electromagnetic/ thermal/mechanical effects for cavity design;

On top of these field solvers, we have built an optimization solver to solve optimization (forward) and shape determination (inverse) problems, and a mesh movement tool to model imperfections. These advanced accelerator modeling tools utilizing massive parallel computing resources have brought about significant advancement to the RF development of accelerator components. In the following sections, we will present the impact of using these codes on various accelerator projects.

3. Numerical Prototyping

The high accurate modeling enabled determining hardware dimensions with accuracies better than machine tolerance. Using these codes, one can produce hardware dimensions directly by the numerical calculations for the engineering drawings, eliminating the conventional trial-and-error prototyping processes used to determine the final machine dimensions.

3.1. *Room temperature structure design for high energy linear colliders*

The damped-detuned accelerator structure for the JLC/NLC [3] was the first structure with dimensions determined directly using the parallel simulation code Omega3P. The JLC/NLC was a technology option using room-temperature RF for a TeV scale e+e- linear collider for HEP research. The Round Damped, Detuned Structure (RDDS) [4] for the NLC main linac was a novel 3D design that is optimized for higher gradient (14% increase over standard design) and suppresses harmful long-range wakefields that disrupt long bunch train operation. The computational challenge was to model the complex geometry close to machining tolerance in order to obtain accuracy for the cavity frequency of better than 0.01%. This is necessary for maintaining the cavity efficiency and to avoid post-tuning of the cells. Using Omega3P on the NERSC's Cray T3E, a table of dimensions for all 206 cells along the RDDS was generated for computerized machining, achieved a modeling accuracy for complex 3D geometries suitable for engineering drawing (1998) — then the only code can achieve such an accuracy. RDDS1 structure was manufactured by KEK with a frequency deviation within 0.5 MHz RMS. This is a significant result which demonstrates that high resolution modeling can replace time-consuming and costly trial-and-error prototyping in the RF structure development. Though the JLC/NLC linear technology has long been replaced by the superconducting technology, the modeling capabilities offered by such tool based on parallel algorithm and supercomputing have been readily applied to many other accelerator R&D, such as the LCLS, ILC, LHC projects.

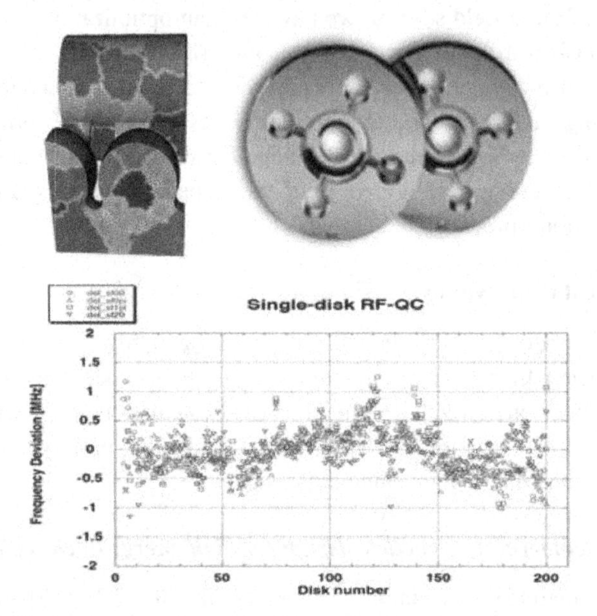

Fig. 2. (Top left) Partitioned model of the RDDS cell; (Top right) Prototypes fabricated with dimensions calculated by Omega3P; (Bottom) QC measurements on 206 DDS cells showing deviation of 1 MHz from targeted frequency of 11.424 GHz.

3.2. *LCLS RF gun design*

The LCLS RF Gun [5,6] was optimized using Omega3P to produce highly cylindrical symmetrical RF fields in a 3D cavity geometry for generating high quality and low emittance beams for the X-Ray FEL. This cavity employs a 3D racetrack dual-feed coupler design to minimize the dipole and quadrupole fields and reduce pulse heating by rounding the z-coupling iris (Fig. 3). The multipole components in concern are many orders of magnitude smaller than the accelerating field, requiring high accurate field calculations to quantify the field quality and to ensure no emittance degradation due to these fields. The RF Gun was manufactured using the dimensions produced by Omega3P. The agreement of the measured RF parameters is in remarkable agreement with the Omega3P calculation: PI-mode frequency in GHz (2.855987 vs 2.855999), Qo (13960 vs 14062), β (2.1 vs 2.03), mode separation in MHz (15 vs 15.17) and field balance (1 vs 1) [7]. The RF Gun was installed in the LCLS injector and was able to deliver high quality beams for the X-ray FEL, meeting the design requirements.

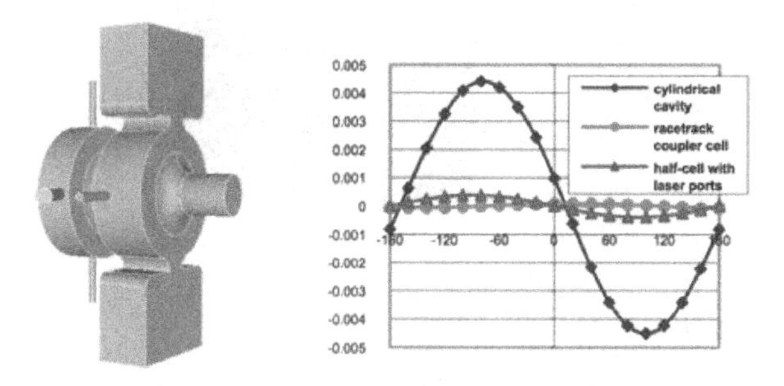

Fig. 3. (Left) LCLS RF gun; (Right) Quadrupole moment in the gun cavity before and after optimization.

The beam dynamics of the LCLS RF Gun was studied using Parmela [8] and IMPACT [9], which was later benchmarked with the Particle-in-Cell (PIC) Pic3P solver of the ACE3P suite (which became available after the RF gun was designed). The Pic3P code is a parallel implementation of the PIC method on an unstructured grid with quadratic conformal elements and adaptive refinement of the basis function order (up to 6th), resulting in a highly efficient code. Pic3P simulates self-consistent charged particle beam formation and transportation in RF circuits such as RF guns and high power klystrons. Unprecedented accuracy can be obtained owing to the higher-order particle-field coupling on unstructured grids and parallel operation on supercomputers. The implementation of a casual moving window further reduces the requirement of computational resources [12]. Fig. 4 shows a snapshot of the scattered self-fields excited by the bunch as it is traversing the RF Gun.

The combination of the high accurate field solver Omega3P and the full wave simulation Pic3P solver provided a powerful parallel simulation tool set for electron source modeling. These codes can be applied to the simulation of large systems such as the entire RF photo-injector and the end-to-end high power klystron, which will allow system scale optimization for beam quality and RF efficiency.

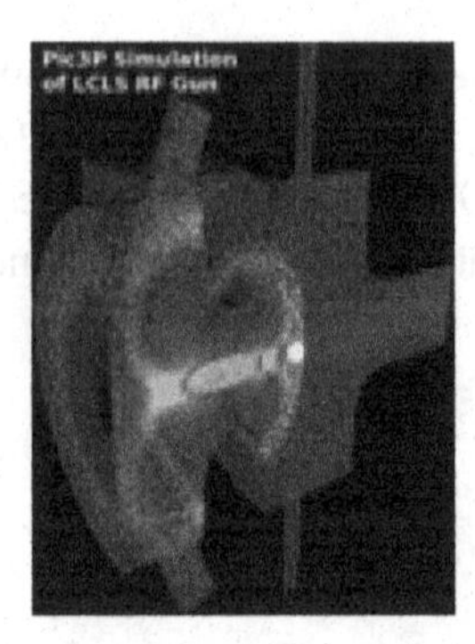

Fig. 4. A snapshot of scattered field in time of a charged bunch traversing the LCLS gun cavity.

3.3. *HOM damping in ILC LL cavity*

The Low-Loss cavity design [10, 11] (Fig. 5) is being considered as a possible upgrade to the baseline TESLA cavity for the ILC main linac. The optimized LL shape has 10% lower B_{peak}/E_{acc} ratio and 15% higher R/Q than the TESLA design. With the low B_{peak}/E_{acc}, the LL design could support a higher field gradient, and reduce the cryogenic loss by 20%. The realization of such a design for the ILC linacs would lead to significant cost saving in machine construction and operation. While the high precision modeling of the geometry dimensions is less critical in the superconducting RF (due to a different machining process), wakefield damping is an important cavity design requirement in order to preserve the beam quality. The wakefield damping optimization requires full cavity (and multi-cavity system) modeling.

The end-group of the LL design was optimized using Omega3P for HOM damping. A high fidelity mesh consisting of 0.53M quadratic elements (3.5M DOFs) was used to model the cavity. This provides sufficient resolution for modifying the end-groups to improve the HOM damping. Figure 5 shows the partitioned mesh used for the parallel calculation of the HOM damping. Figure 6 shows the damping results of the dipole HOMs.

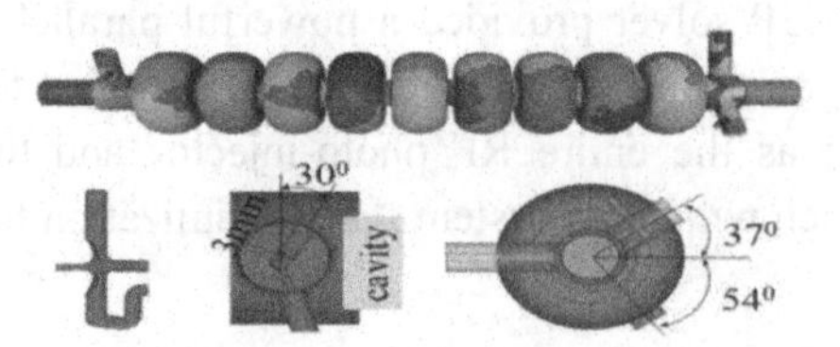

Fig. 5. Partitioned mesh and optimized end-groups of the LL shape cavity for effective HOM damping.

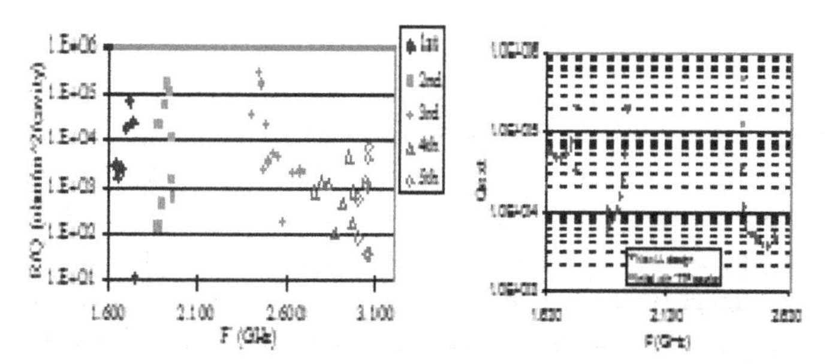

Fig. 6. Dipole mode R/Q and damping results of the LL cavity using Omega3P.

4. Numerical Experiment and Diagnostics

A cavity design can be evaluated using the ACE3P solvers by virtual numerical experiments — to simulate the RF systems under realistic "experiment" conditions. The simulation results can in turn provide guidance on improvements of the design before the hardware is being manufactured. Of particular interests for numerical studies in superconducting RF are the cavity imperfection and multipacting.

Imperfections in superconducting cavities may be induced in the manufacturing process, the tuning and the handling. Understanding the tolerance requirements on cavity imperfection is important for the SRF cavity development. We have developed a mesh deformation tool as part of the ACE3P suite. One can use the mesh deformation tool to generate a set of deformed mesh models of the cavity with predefined cavity imperfection types. Then use the Omega3P solver to analyze the effects of imperfections on RF and wakefields/beam dynamics, providing understanding of tolerances of cavity geometry. Further, the multi-parameter optimization solver of ACE3P can also be used to solve inverse problems using the experimental data as input to discover the actual imperfection of the cavity.

Multipacting (MP) is an issue of concern for superconducting resonators which may cause prolonged processing time or limit the achievable design gradient. While most of the MP bands may be conditioned and eliminated with RF, hard multipacting barriers may prevent the resonators from reaching the design voltage. Elimination of potential MP conditions in the cavity design could significantly reduce time and cost of conditioning and commissioning.

ACE3P provides a set of tools to perform "numerical experiments" on cavity designs to evaluate and validate the RF performance.

4.1. *Imperfection study of the ILC TDR cavity*

In the 9-cell TDR cavity, each dipole band consists of 9 pairs of modes. The degeneracy of the pairs is split by 3D effects of the couplers. With cavity imperfection, the frequencies and Q factors of different cavities scatter and shift from the ideal cavity. Figure 7 shows the frequencies and Qe's of the 6th pair in the second dipole band of the 8 cavities in a TTF module [12]. The two black dots are the Omega3P results for the ideal cavity. Compared with the ideal case, the splitting of the measured mode pairs is larger, the mode pair mostly shifts to lower frequencies, and the Qe's scatter mostly toward high values. The Qe increase would be problematic if their values exceed the beam stability limit. The large frequency split of the mode pair is the result of perturbation to the field distribution by the imperfection, which may induce x-y coupling and dilute the beam emittance.

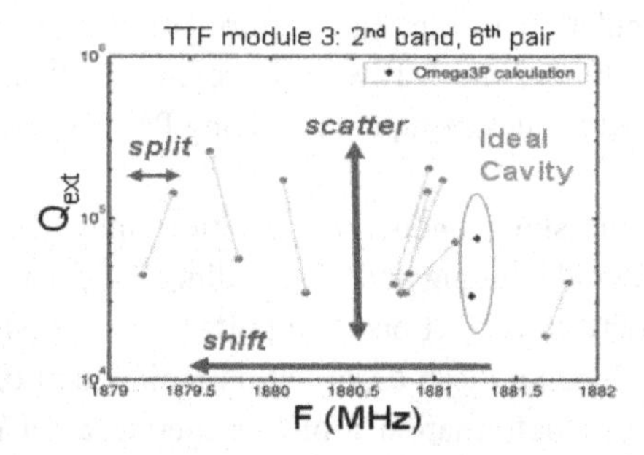

Fig. 7. Scatter in measured dipole mode data in TTF cavities.

In order to model realistic cavities for the ILC beam dynamics study, we have embarked on an effort to determine the true cavity shape by solving an inverse problem [13, 14], using the data from the TESLA test data as input parameters. Progress has been made in identifying key imperfection parameters that cause the discrepancy between the measured data and the Omega3P calculation of an ideal cavity. Figure 8 shows some possible cavity deformations types identified through the shape determination process. Figure 9 shows the polarizations of the 6th dipole pair in the 2nd dipole band for the ideal cavity and a deformed cavity with elliptical cells. It can be seen that the polarizations of the dipole pair are quite different for the two cases, and their changes in the deformed cavity could lead to the enhancement of x-y coupling for wakefield effects. These deformation parameters can be used to produce realistic cavities using the mesh deformation tool for Omega3P calculations. The RF parameters of the deformed cavities can then be provided to the beam dynamics studies.

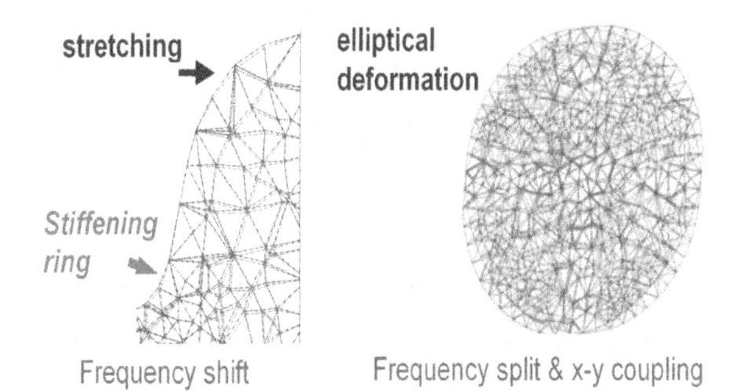

Fig. 8. (Left) Deformed cell surfaces that cause frequency shift; (Right) Elliptical cell deformation that causes frequency split as well as x-y coupling.

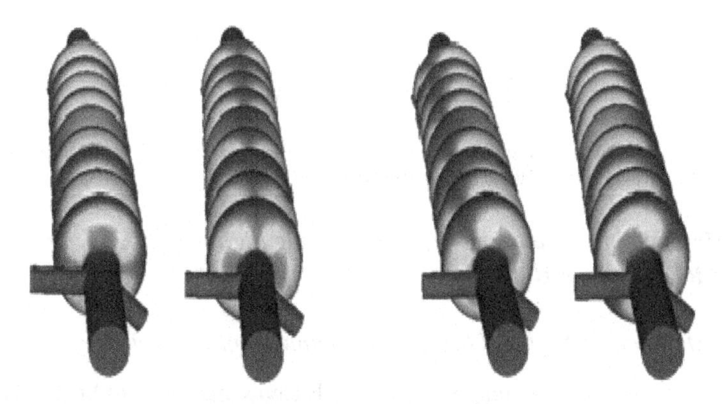

Fig. 9. Field patterns of a dipole pair in (Left) ideal cavity; (Right) cavity with elliptical cells.

4.2. *Diagnostics — high Q mode that cause CEBAF BBU*

In the CEBAF 12-GeV upgrade, beam breakup (BBU) was observed at beam currents well below the design threshold. Using measured RF parameters such as frequency, Qext, and mode field profile as inputs, the solutions to the inverse problem identified the main cause of the problem by recovering the true shape of the cavity [15]. Due to fabrication errors, the cavity was 8 mm short as predicted and confirmed later from measurements (Fig. 10). The 8-mm cavity shortage has a particular distribution that caused the three HOMs in question to have fields shifted away from the HOM coupler region which resulted in ineffective damping (Fig. 11). The effort to resolve the CEBAF puzzle showed that experimental diagnosis, advanced computing and applied mathematics working together solved a real world problem as intended by SciDAC.

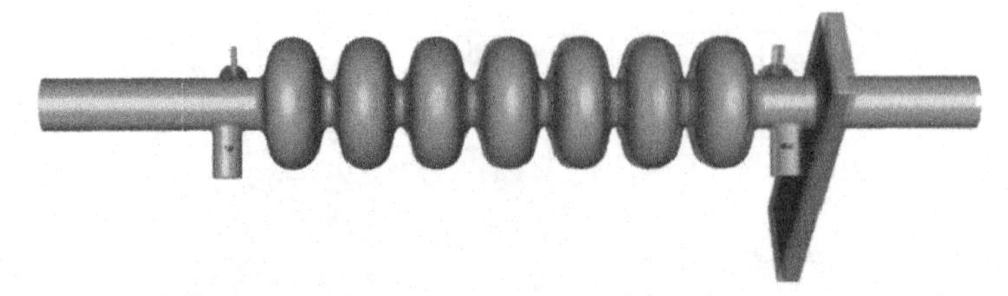

Fig. 10. The CEBAF SRF original (silver) and deformed (gold) cavities.

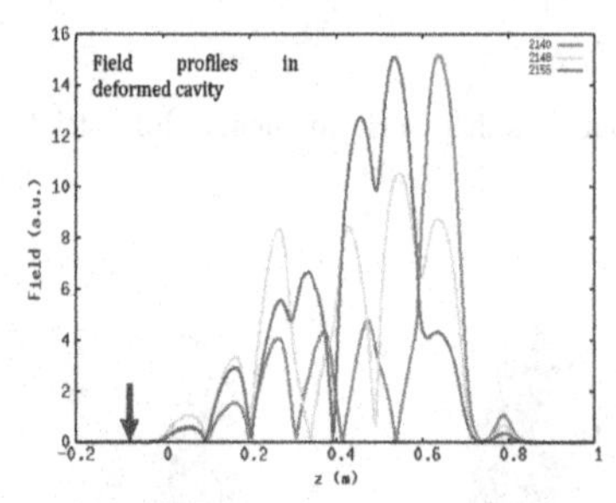

Fig. 11. The fields of three problematic modes found in one of the CEBAF upgrade prototype cavities were shifted from the HOM coupler, causing abnormally Qext.

4.3. *Multipacting simulation — understanding high power behavior of RF components*

4.3.1. *MP analysis for the ILC superconducting cavity RF coupler*

Couplers are important components of an RF cavity and need to be conditioned for high power handling, which sometimes can be a long process. The TTF-III coupler [16] (Fig. 7) adopted for the ILC baseline cavity design, as an example, has shown a tendency to have long initial high power processing times. A possible cause for the long processing times is believed to be multipacting in various regions of the coupler. To understand performance limitations during high power processing, SLAC has built a flexible high-power coupler test stand [17]. The plan was to test individual sections of the coupler using the test stand to identify problematic regions. To provide insights to the high power test results, detailed numerical simulations of multipacting for these sections have been carried out using Track3P. Figure 12 shows a model of the "cold" coax in the high power test. In addition to monitoring the vacuum pressure, an electron pickup was placed on the outside wall of the coax to measure the electron activities. The calculated MP bands as well as the measured pickup signal and

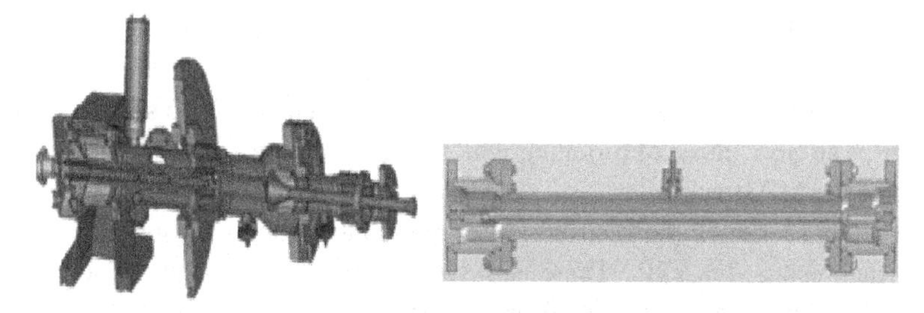

Fig. 12. (Left) the TTF-III coupler; (Right) cold coax test setup.

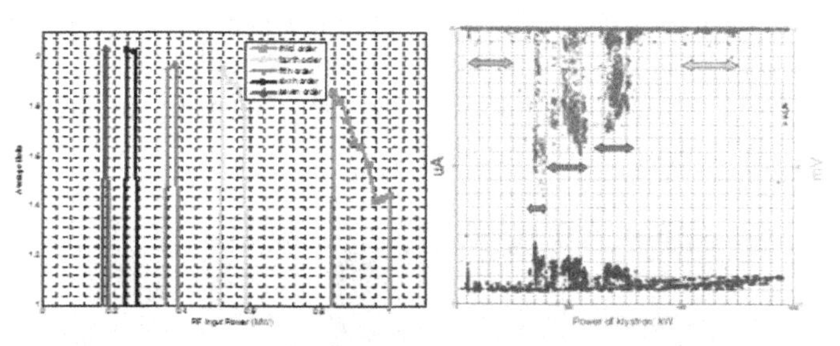

Fig. 13. (Left) MP bands from Track3P simulation and (Right) high power measurement data of the cold coax.

the vacuum current after initial processing showing in Fig. 13 exhibit good agreement between simulations and measurements [17, 18].

4.3.2. *MP analysis for the FRIB β=0.085 quarter wave resonator*

The driver linac for the Facility for Rare Isotope Beams (FRIB) [19] utilizes several types of low-β superconducting resonators to accelerate the ion beams from 0.3MeV per nucleon to 200MeV per nucleon. Multipacting (MP) is an issue of concern for such superconducting resonators as they have unconventional shapes. We have used the parallel codes Track3P and Omega3P to analyze the multipacting barriers of such resonators.

Figure 14 shows four snapshots of the Track3P simulation of the β(v/c) = 0.085 QWR cavity [20]. Seed particles were initiated on all the RF surfaces. The field level was scanned up to 1.5MV accelerating voltage with a 0.01MV interval. At each field level, 50 RF cycles were simulated for obtaining resonant trajectories. Several snapshots of particle tracking are shown in Fig. 2 to illustrate MP resonances at 23kV accelerating voltage. The primary particles were emitted from all the surfaces (a); after a few RF cycles, some non-resonant particles were removed from the simulation (b); after more RF cycles, fewer particles survived (c); finally, after 50 RF cycles, only resonant particles which are considered potential MP particles remained (d). Both 2eV and 5eV were used as the initial energy for primary and secondary emissions to study its effect on MP. The SEY curve of niobium, which peaks at around 350 eV, was used to estimate the MP strength.

Figure 15 shows the distribution of resonant particles versus gap voltage identified using Track3P. There are two potential MP bands, one at low fieldlevels with impact energies from tens of eV to about 2keV, and the other at high field levels from 360kV to 600kV accelerating voltage are with low impact energies. In RF testing of the QWRs, several low and medium field MP barriers were encountered in the Dewar tests of 5 β=0.085 QWRs [20]. Those barriers appeared to be difficult to processed through. Geometry modifications were considered to mitigate the MP problems.

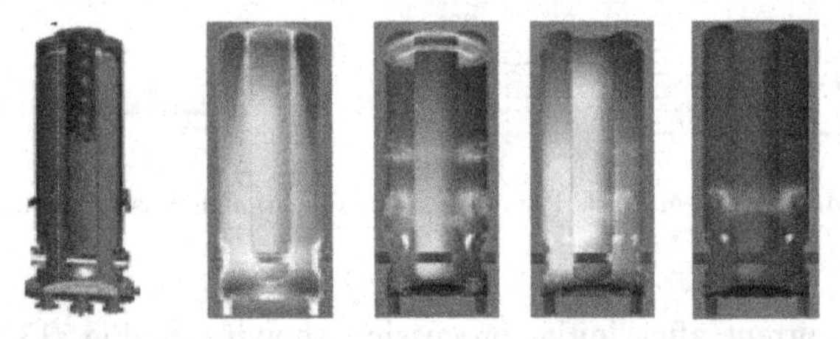

Fig. 14. (Left) MP simulation model. (Right four) Evolution of particles survived at increasing RF cycles. Particles survived a large number of RF cycles are considered potential MP particles.

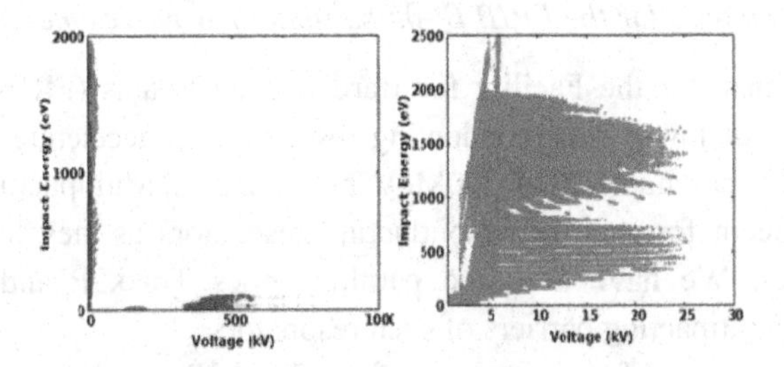

Fig. 15. Multipacting activities at low accelerating voltage with initial particle energies 2eV. Left: Impact energies vs. accelerating voltage. Right: zoom in view of multipacting at low voltages.

5. Large Scale Simulation — System Diagnostics and Validation

Large scale simulation allows users to simulate a complete RF system or unit to validate the design concepts.

5.1. *ILC cryomodule simulation*

An ILC RF unit consists of three cryomodules driven by one klystron, and each cryomodule has 8 or 9 cavities. While HOM damping has been optimized for individual cavities, its effects for modes above the beam-pipe cutoff have not been fully investigated. It is of great interest to understand the trapped modes and their damping in the RF unit with realistic cavity imperfections and misalignments, to study beam heating and the effectiveness of beamline absorbers in the interconnects between cavities and cryomodules. Successful simulations of this large system would provide invaluable information on issues that are important to the machine performance, enabling system validation in the R&D phase. As a first step toward this goal, the first-ever calculation of dipole modes in a cryomodule has been successfully performed on the NERSC supercomputer. A typical mode in the 3rd dipole band is shown in Fig. 16. One can see that the polarization of the mode rotates through the cryomodule. The effects could induce significant x-y coupling to beam dynamics, which need to be understood and minimized. The effectiveness of HOM couplers on damping these modes is determined by the field patterns in the interconnects, and will be strongly affected by cavity imperfections and misalignments. These issues are best addressed through simulating the multi-cavity system.

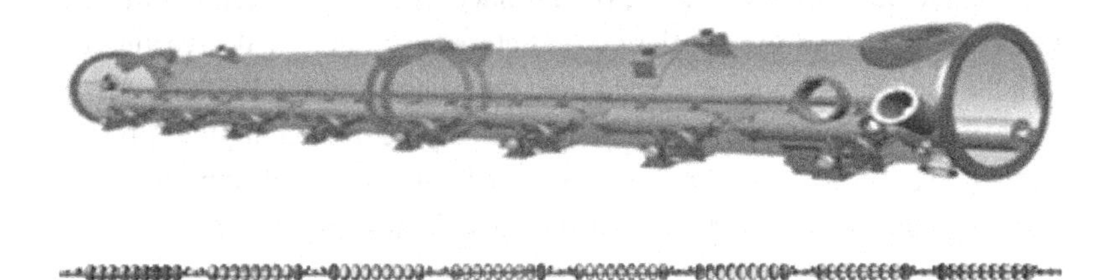

Fig. 16. ILC 8-cavity TTF cryomodule and a HOM mode in the 8-cavity module from Omega3P.

In simulating the eight-cavity cryomodule on one of NERSC's older computers (Seaborg), the runtime was 1 hour per mode using 300 GB of memory on 1024 CPUs for the problem size with 20 million DOFs. It is estimated that 200 million DOFs would be required to simulate the ILC 3-cryomodule RF unit. Petascale computing resources together with advances in computational science research will definitely be needed in achieving this goal. In fact, in order to successfully simulate the above cryomodule on current supercomputers, a new memory reduction scheme was developed for the parallel linear solver that enables to solve problems twice as large with the same available memory [21].

5.2. *CLIC two-beam accelerator — power transfer and wakefield coupling*

The Compact Linear Collider (CLIC) [22] provides a path to a multi-TeV accelerator to explore the energy frontier of High Energy Physics. Its two-beam accelerator (TBA) concept envisions complex 3D structures, which must be modeled to high accuracy so that simulation results can be directly used to prepare CAD drawings for machining. The required simulations include not only the fundamental mode properties of the accelerating structures but also the Power Extraction and Transfer Structure (PETS), as well as the coupling between the two systems. Applying SLAC's parallel finite element code suite on largest supercomputers available, for the first time, power transfer and dipole wakefield coupling effects were computed for the fully 3D combined system of one PETS, two TD26 accelerator structures and a connecting waveguide networks [23].

Figure 17 (Right) shows a temporal snapshot of fields generated by a multi-beam bunch train in the PETS structure propagating toward the accelerating structure. The left plot in Fig. 18 shows the accelerating voltage in the accelerating structure by the power produced in the PETS. The red is the voltage by the power of a single bunch driving the PETS and the green is the voltage due to the power of a bunch train in the PETS.

The waveguide network can also produce coupling of dipole wakefields between the PETS and the accelerator structure. For numerical investigation of potentially dangerous transverse wakefield coupling from the PETS to the AS, a T3P simulation of wakefields excited by a dipole bunch in the PETS was performed. The right plot in Fig. 18 shows the coupled longitudinal and transverse dipole wake potentials in the two accelerator structures.

Fig. 17. (Left) CLCI two beam accelerator unit. (Right) A snapshot T3P simulation of power transfer from PETS to the accelerator structure.

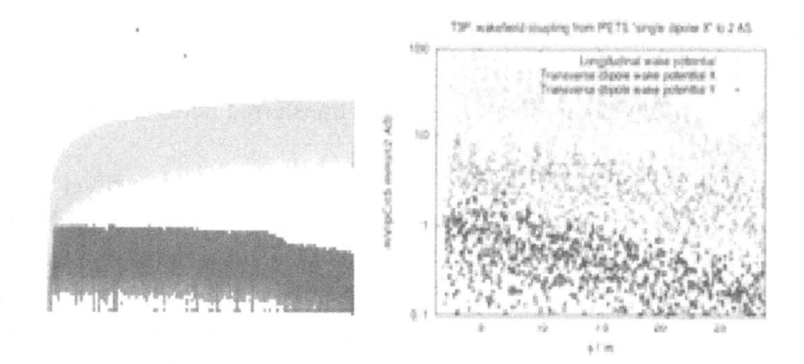

Fig. 18. (Left) Accelerating voltage in the accelerator structure due to the power generated in the PETS of a single bunch (red) and multi-bunch train (green). (Right) Wakefield coupled from the PETS to the accelerator structure.

6. Multiphysics Modeling — Seamless RF and Engineering Design

The multi-physics modeling toolset of ACE3P includes the RF, thermal and mechanical solvers. These solvers utilize the same data structure and user interface, enables the users to perform seamless RF & engineering design and analysis.

The thermal and mechanical solver TEM3P compliments the existing RF finite-element code developed at SLAC and is built upon the same code infrastructure as the RF solvers, such as Omega3P and S3P. The solver provides capabilities of multi-physics analysis of RF heating, structural effects, mechanical vibration and Lorentz Force Detuning in addition to the RF capabilities provided by Omega3P and S3P solvers. The ACE3P code suite provides a complete analysis toolset for engineering prototyping in a single development framework. Parallel implementation in TEM3P allows large-scale computations on massively parallel supercomputers with a fast turnaround time. The multi-physics simulations is done in four steps: (1) electromagnetic simulation for the vacuum region; (2) thermal simulation for the cavity metal body; (3) mechanical simulation for the cavity metal body; (4) calculation of thermal frequency drift caused by structural deformations. Here we provide the multi=physics analysis of the LCLS RF Gun and a superconducting cavity HOM coupler to illustrate of the use of these solvers.

6.1. *Multi-physics analysis of LCLS RF gun*

The analysis starts from a CAD model of the RF gun shown in Fig. 19. Second-order finite element meshes are generated using CUBIT for the vacuum and metal body regions of the RF gun. The EM simulation was applied to the vacuum region of the mesh and the thermal and mechanical analyses to the metal region. The computational data are transferred between different analyses through a common data structure.

The electromagnetic analysis performed using the parallel eigensolver Omega3P determines the resonant mode frequencies and field distributions. The heat source for the thermal simulation is the power loss of the accelerating mode on the cavity wall. The numerical linear system of equations of the thermal solver is solved using an iterative solver. Figure 19 illustrates the simulation cycle performed for a RF thermal and mechanical multi-physics analysis of the LCLS RF Gun.

6.2. *Superconducting cavity HOM coupler*

The TEM3P multi-physics solver is capable of solving nonlinear thermal and structural problems, such as those arising from analysis in superconducting cavities. Under superconducting environment, the material properties have strong dependences on temperature and in most cases non-linear. Realistic material properties can be provided to the multi-physics solver via data points or polynomial functions. The solver employs an iterative approach to obtain a self-consistent and converged solution. Figure 20 shows a plot of the temperatutre distribution of the JLAB HCCM cavity HOM coupler calculated by TEM3P [24]. The temperature on the cavity-coupler body ranges from 2K to 300K, subject to a wide range variation of thermal and mechanical properties of the cavity-coupler body materials. Careful design and accurate evaluation of the temperature map is important to ensure the cavity perform as designed.

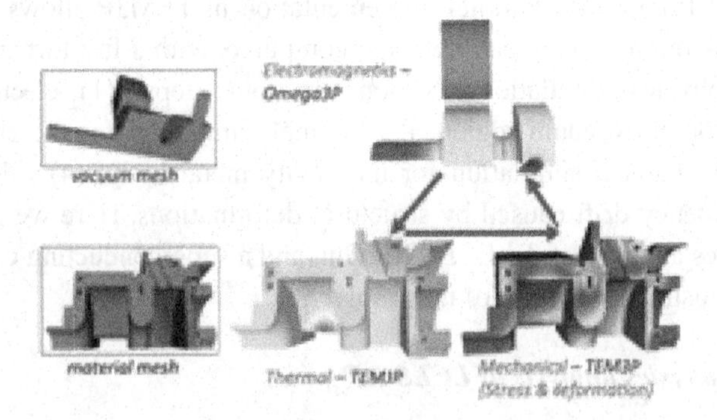

Fig. 19. RF, thermal, and mechanical simulation models and a full cycle of multi-physics analysis for the LCLS RF Gun.

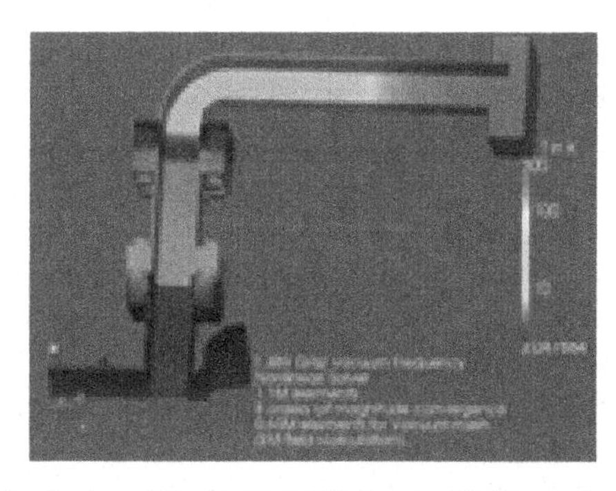

Fig. 20. Temperature distribution of the JLAB HCCM cavity HOM coupler calculated by TEM3P.

7. Final Remarks

Advanced computing has become an indispensable part of research and development of modern accelerators and light sources. The high fidelity modeling capabilities offered by the advanced modeling codes not only provided a set of powerful tools for the design and optimization of RF components, but also a set of comprehensive tools for analyzing system performances under realistic operating conditions and tolerances. With the ever increasing computing power offered by modern supercomputers and new functionalities and capabilities integrated into these toolsets, advanced modeling will have a greater impact on the way accelerators are being designed and operated.

Acknowledgments

The author would like to thank colleagues in the Advanced Computations Department for the materials presented in this paper. And would like to thank our colleagues at various accelerator laboratories for collaborations on accelerator applications, and various SciDAC CETs and Institutes for collaborations on computational science. The work was supported by the U.S. DOE contract DE-AC02-76SF00515. The work used the resources of NERSC at LBNL which is supported by the Office of Science of the U.S. DOE under Contract No. DE-AC03-76SF00098.

References

1. http://www.scidac.gov/.
2. Kwok Ko, *et al.*, "Advances in Parallel Electromagnetic Codes for Accelerator Science and Development," proc LINAC10, Tsukuba, Japan, 2010.
3. http://www.linearcollider.org.

4. Z. Li *et al.*, "Design of the jlc/nlc rdds structure using parallel Eigensolver omega3p," proc. PAC1999.
5. https://portal.slac.stanford.edu/sites/lcls_public/Pages
6. L. Xiao *et al.*, Dual Feed RF Gun Design for the LCLS, Proc. PAC 2005, Knoxville, Tennessee, May 15–20, 2005.
7. D.H. Dowell *et al.*, "Results of the SLAC LCLS Gun High-Power RF Tests," proc. PAC2007.
8. http://laacg.lanl.gov/laacg/services/serv_codes.phtml
9. J. QIang *et al.*, Recent Improvements to the IMPACT-T Parallel Particle Tracking Code," proc. ICAP 2006.
10. J. Sekutowicz *et al.*, JLAB, TN-02-023, June 2002.
11. Z. Li *et al.*, "Optimization of the Low Loss SRF Cavity For the ILC", proc. PAC2007.
12. L. Xiao *et al.*, "Modeling Imperfection Effects on Dipole Modes in TESLA Cavity", proc. PAC2007.
13. L. Lee *et al.*, "Modeling RF Cavity with External Coupling", 2005 SIAM Conference on Computational Science and Engineering, Orlando, Florida, 2005.
14. V. Akcelik *et al.*, "Adjoint Methods for Electroma gnetic Shape Optimization of the Low-Loss Cavity for the International Linear Collider", Proc. SciDAC 2005, San Francisco, California, 2005.
15. Z. Li *et al.*, "Analysis of the Cause of High External Q Modes in the JLab High Gradient Prototype Cryomodule *Renascence*," SLAC-PUB-13266, June 12, 2008.
16. http://tesla-new.desy.de
17. Brian Rusnak, *et al.*, "High-Power Coupler Component TestStand Status and Results", these proceedings.
18. L. Ge, *et al.*, "Multipacting Simulations of TTF-III Coupler Components", these proceedings.
19. http://www.frib.msu.edu
20. L. Ge *et al.*, "multipacting simulation and analysis for the frib β=0.085 quarter wave resonators using track3p," proc. IPAC2012.
21. L.-Q. Lee *et al.*, "Enabling Technologies for Petascale Electromagnetic Simulations", Proc. SciDAC 2007, Boston, Massachusetts.
22. http://clic-study.org
23. A. Candel *et al.*, "Numerical Verification of the Power Transfer and Wakefield Coupling in the CLIC Two-beam Accelerator," SLAC-PUB-13329.
24. V. Akcelik *et al.*, "Thermal Analysis of SRF Cavity Couplers Using Parallel Multiphysics Tool TEM3P," SLAC-PUB-13638, 2009.

Transverse Wakefields and Means to Suppress Wakefields in High Gradient Linear Colliders

Roger M. Jones

School of Physics and Astronomy, The University of Manchester, Oxford Road, Manchester, M13 9PL, UK

Cockcroft Institute of Science and Technology, Daresbury, WA4 4AD, UK

The symposium held at Tsinghua University, in honor of Dr. Juwen Wang, is a testament to his engagement with diverse areas of linear accelerators. My work has overlapped with his during the period of the Next Linear Collider/Japanese linear Collider (NLC/JLC) programe in particular. Here I report on the beam-exited long-range wakefield in these linacs, and also on subsequent developments in the context of the Compact Linear Collider programme (CLIC). This self-excited wakefield, if left unchecked, can at the very least cause a marked dilution in the beam emittance, and in the worst case can cause a catastrophic beam break up (BBU) instability rendering the particle beam unusable. To ameliorate this affect, there are two main strategies: one can either aim at heavily damping the wakefield (in practise requiring a Q value as low as ~10) or one can detune each of the accelerator's cells to ensure that each mode is excited at slightly different frequency, in a precise manner. The former approach entails placing damping waveguides and damping materials in relatively close proximity to the beam, and this is the approach adopted by the CLIC collaboration. For the NLC/JLC we collectively followed the latter approach, in which the dipole mode of each cell is detuned with an erf function profile along each accelerator structure. Eventually the modes, which form the wakefield, recohere and, to ensure the wakefield remains below a specified level a portion is coupled out through slots cut into each cell to an attached waveguide-like manifold. Typically medium Q values are aimed at, between 500 and 1000. This scheme entails suppressing the modes which comprise the wakefield –using damped and detuned structures (DDS). Sampling an attenuated portion of this manifold radiation also provides both a beam and structure diagnostic. A similar DDS design, but with more stringent wakefield suppression requirements, has been followed as an alternative to the CLIC damping scheme. Wakefield suppression for the NLC/JLC and CLIC is reported on, with a focus on the DDS approach.

1. Introduction to High Gradient Linacs in Linear Colliders

The largest particle accelerator at present is the LHC [1,2], which is a circular collider of circumference 27 km, with a design goal of colliding hadrons at a

14 TeV centre of mass in 2015. Plans are already being made for a successor, which may also be another circular collider, but at an increased circumference of 100 km. This scenario, known as the Future Circular Collider (FCC), aims at colliding hadrons at a centre of mass energy of 100 TeV. In addition, there are also plans for another, scheme known as the compact linear collider (CLIC) which, as synchrotron radiation is not an issue, accelerates and collides leptons. Electron-positron collisions are hence facilitated by a normal conducting (NC) linear collider. A lepton collider will enable precision collisions and in this manner is complementary to a hadron collider. Acceleration of either single or multiple bunch can be envisaged. The former design was proposed in the VLEPP [3] scheme. The focus of current research efforts for CLIC is however on multi-bunch acceleration. In order to reach a centre of mass collision energy of 3 TeV high gradient accelerating linacs are required. A staged design, at lower energies is also under consideration –at 500 GeV and 1.5 TeV [4]. Lower energies, of up to 1 TeV are also within the reach of superconducting (SC) technology. The international linear collider (ILC [5–6]) design is focused on a 1TeV energy with a gradient of 31.5 MV/m, along a linac which is approximately 10 km long. Each SC cavity, based on the TESLA technology choice [7], consists of 9 cells and, depending on the location within the linac, there are 8 or 12 of these cavities within a given SC module. There are also alternate designs which aim at a gradient of 50 MV/m. However, apart from single cells [8–9] demonstrating the technique experimentally, no complete cavity structures have attained this gradient –although there is an international campaign in progress to do so. High gradients are within the reach of NC cavities operating at room temperature. This paper is focused on high gradient NC accelerators.

Up until 2004 there were two main NC linear collider research efforts. These were concentrated on the Next Linear Collider/Global Linear Collider (NLC/GLC [10–12]) design and on the Compact Linear Collider (CLIC) design [13–15]. After careful consideration, the International Technical Recommendation Panel (ITRP) decided to consolidate the efforts of the SC TESLA and NC NLC/JLC research efforts into a single SC technology choice. The CLIC NC design nonetheless remains an option to reach high centre of mass energies and this is achievable through the larger gradient achievable in NC accelerator structures. Indeed, without the requisite accelerating gradients the overall footprint of the collider would be prohibitively large. The present baseline design for CLIC incorporates a centre of mass energy of 3 TeV with an average loaded accelerating gradient of 100 MV/m (although there is also a design for an electron-positron collision scheme for CLIC at 500 GeV). By necessity, NC accelerators for linear colliders must operate in a short pulse mode, otherwise Ohmic losses in the walls of the cavity will result in a prohibitively low overall efficiency. Whereas for SC accelerators, a long pulse train is acceptable as there are minimal RF losses. The ILC bunch train is approximately 1 ms long and the CLIC bunch train is markedly shorter at 156 ns. The fundamental difference of the CLIC scheme from the ILC lays in the dual-beam method used in CLIC. The drive beam operates at a high current of 100A and is decelerated in order to serve as a source of RF power to the main accelerating beam which operates at a current of 1A. The reason for utilising a two-beam scheme design comes from the necessity of reducing the number of rf klystrons to a manageable number in a 3 TeV lepton collider. The detailed components and parameters in the CLIC scheme are discussed in [12]. We confine this work to the e.m. field excited in the cavities of each of the main accelerating linacs, each of which is 21.1 km long and will contain more than ~71,000 structures [16]. The wealth of knowledge gained in more than two decades of research on high gradient cavities at SLAC in particular is a valuable resource for CLIC as rf breakdown and wakefield suppression issues, for example, are common concerns for both NLC/GLC and CLIC.

In general, the introduction of a charged particle bunch into a cavity excites an electromagnetic (e.m.) field which can be decomposed into an infinite series modes. It also often the case, that a limited number of modes have the most severe impact on the particle beam dynamics. This e.m. field acts both on the bunch itself and on bunches within the train. The e.m. field is conveniently represented as a wakefield [17–18], in which we refer to that acting along the bunch as the short-range wakefield and that further along the train as the long-range wakefield. Furthermore, the wakefield can be decomposed into that acting both longitudinally and transverse to the acceleration axis. The former modifies the bunch energy and can be controlled by BNS [19] damping and the latter modifies the bunch emittance and can cause a severe BBU instability [20]. In this work we focus on means of suppressing the transverse long-range wakefield. For single bunch linear colliders, such as that envisaged in the VLEPP design operated at high repetition rates, long-range wakes are of course not an issue. However, the transverse wakefield is severe for the accelerating structures designed for multi-bunch machine CLIC. Also, the average iris radius is considerably smaller than that in the ILC and hence the wakefields are significantly larger. In the diffraction limit, valid for short bunches, the envelope of the monopole (dipole) longitudinal (transverse) wakefield is inversely proportional to $a^{1.7}$ ($a^{2.7}$) [17, 21], where a is the average cell radius of an accelerating cavity. In general, longitudinal and transverse wakes, from geometrical scaling considerations, can also be shown to be proportional to frequency as ω^2 and ω^3, respectively. The average iris radius of cells in the ILC cavities is approximately 35 mm and this compares with 3 mm for CLIC cavities. The CLIC accelerating frequency is 12 GHz and this is almost an order of magnitude larger than the ILC monopole frequency of 1.3 GHz. The CLIC accelerating structure X-band frequency is the result of an optimization [22] entailing maximizing beam luminosity whilst minimizing the cost and at the same time ensuring rf breakdown constraints (minimization of surface e.m. fields in particular), are adhered to. The focus of this paper is on means of suppressing the wakefield in NLC/GLC structures at X-band frequencies (although C-band and S-band NC accelerators have also shown promise as potential candidates for next generation high energy colliders).

This paper is organized such that the next main section provides an overview of wakefield suppression in NC structures. The next main section discusses damped and detuned means of wakefield suppression in NC accelerating structures for the NLC/GLC programme and some initial results on a structure for the CLIC project. This concludes with a brief discussion on the implications on beam dynamics.

2. Overview of Wakefield Suppression in NC Structures

2.1. *Introduction*

There are several possible methods available to suppress transverse wakefields excited in multi-cell cavities of a NC linear collider. However, it is important that in suppressing the parasitic modes excited by the beam the accelerating mode is left largely unaffected. There are several approaches available to affect this wakefield suppression. These methods can be grouped in two main schemes: heavy damping (which entails Q values as low as ~10) [23] and moderate damping (with Q values ~300–1000). The former method is the baseline design for the CLIC main linac accelerating cavities and entails waveguide-like structures directly impinging into each accelerating cell. The latter method relies on heavy detuning of the cell frequencies and often takes advantage of the natural partitioning of modes into characteristic pass-bands (in the structures with a $2\pi/3$ per cell phase advance the modes which dominate the transverse momentum kick are located in the first pass-band [24–25]) and hence only a limited number of bands need to be damped.

In all long-range wakefield suppression schemes it is important to be able to verify the efficacy of the method employed and indeed to ascertain the accuracy of the simulation tools used. The transverse wakefield can be measured by several means. The most direct method entails ascertaining the transverse momentum kick imparted to trailing bunches. Beam-based measurement is necessarily expensive and requires high precision measurement techniques. Nonetheless, at the SLAC National Accelerator Laboratory there existed a facility, known as ASSET [26–27], which provided just such a measurement system, based on a positron excitation bunch and a measurement of the kick imparted to a trailing electron, or witness, bunch. On the other hand, as electron or positron bunches rapidly become ultra-relativistic (i.e. essentially travelling at the speed of light), the e.m. field is transverse to the direction of motion and is a TEM-like e.m. wave. For this reason, bench measurements, which simulate a charge particle beam using a wire offset from the centre of an accelerating cavity, can be used to measure the beam impedance. This stretched wire measurement technique was originally proposed by Sands and Rees [28]. Taking the Fourier transform of beam impedance, enables the wakefield to be recovered. However, it is often the case that, a large fraction of the field at a particular frequency, is located in a small number of cells of the accelerator cavity. These are known as trapped modes and are often difficult to measure by wire based techniques. Nonetheless, this is a valuable technique to characterize both accelerating structures and various associated components [29–31]. Beam-based measurements do not suffer from these inadequacies. Beam-based measurements at SLAC have been made on the wakefield excited in C-band [32–35], X-band [23, 39–42] and K_a-band [43] accelerating structures.

Heavy damping [23] and, moderate damping together with strong detuning in the cell frequencies are two approaches to wakefield suppression. The focus of this paper is on the latter method and in this case the mode frequencies can be arranged to either invoke non-resonant or resonant suppression.

2.2. *Resonant and non-resonant wakefield suppression with moderate damping*

The wakefield excites a series of modes and the overall wakefield can be represented as a sum over these modes, which have a sinusoidal dependence [17]. One means of resonant suppression entails ensuring the bunches in the train are located at the position of the zeros in the wakefield. The method is known as the zero-point crossing approach to minimizing the effect of the wakefield on beam dynamics. A further method entails ensuring the difference in two dipole frequencies excited are an integer multiple of the bunch repetition frequency. Both of these methods resonantly suppress the wakefield and, will be sensitive to manufacturing errors, introduced inevitably in fabricating several thousand of these accelerating structures. If this scheme is adopted, a careful study of the beam dynamics is required, with expected experimental errors included in the simulations, to provide confidence that the wakefield impacts the emittance [43] and energy of the beam in a well-contained and limited manner.

On the other hand, non-resonant wakefield suppression entails detuning the dipole frequencies excited by the beam and this is the method that has been extensively investigated in the context of the X-band NLC/GLC structures, for the S-Band Linear Collider (SBLC) [46–48] and for the S-band Stanford Linear Accelerator (SLAC) [49]. The advantage of non-resonant wakefield over resonant suppression is that it does not freeze the bunch spacing to a particular value. In essence, non-resonant suppression forces the wakefield to decay for all bunch locations. This method of wakefield suppression can be understood by considering the momentum kick imparted transversely to the acceleration beam axis and this is governed by the transverse kick factor [15]. In forcing the wakefield to decay with a prescribed functional dependence we begin by assigning a specific dependence to the kick factor weighted density function with respect to frequency, Kdn/df. The inverse Fourier transform of Kdn/df gives the wakefield experienced by the initial series of bunches in the train.

The original method of detuning the cell frequencies of each accelerating structure was pioneered at the SLAC National Accelerator Centre. Here, the linacs developed in the 1960s, were intended to accelerate electrons up to 40 GeV through constant gradient accelerator structures operated at a frequency of ~ 3 GHz. However, during initial operation in the late 1960s a BBU instability [20] was observed due to transverse deflecting modes being excited at the same frequency within each cavity. To rectify this, both the focusing quadrupole magnets and the frequencies of the upstream cells were modified [27, 50]. These modification succeeded in enabled the threshold current to be raised to ~60 mA from an initial BBU-limited value of ~20 mA. Changing the cell dimensions of each accelerating structure prevented the modes adding coherently. In this manner the overall transverse momentum kick presented to the beam was diminished. This also had consequences on the overall energy gain –which was reduced by ~1% of the total available beam energy [51].

Similar detuning schemes were applied at S-band for the SBLC project, which aimed at a center of mass collisional energy of 0.5 TeV. Gaussian detuning of the 180-cell accelerating structures was investigated as a means of enforcing wake function suppression using the method outlined in Appendix A. However, the final design dispensed with explicit detuning, as a constant gradient design was chosen (in which the modes will consequently be detuned with constant mode spacing. Over the length of the linacs, ten different structure types were envisaged, each one having a different frequency shift in the dipole modes. In this manner section-to-section detuning of constant gradient structures ensures coherent transverse momentum kicks were minimized and hence the influence on emittance dilution is reduced. This method was adopted in preference to enforcing an explicit Gaussian detuning distribution as it was simpler to implement. In addition, the tips of the irises of each of the cells were coated with a lossy material [27] which ensured the dipole mode Q was kept below 4000. Two sets of HOM couplers, each provided with four orthogonal waveguides, were used to monitor the radiation and from these signals structure alignment could be remotely measured. Each pair of facing waveguides facilitates the monitoring of one particular polarization. This work was abandoned at DESY laboratory, in favor of focused research on a design which utilized superconducting cavities, known as TESLA [7];this technology was later chosen for the ILC [6]. Based on this earlier work, the NLC/GLC design for wakefield suppression was applied to acceleration at an X-band frequency of 11.424 GHz, conveniently chosen as a fourth harmonic of the SLAC linacs. Here an explicit function dependence on the wakefield utilized by carefully tapering the cell dimensions. This approach is described in the next section.

3. Wakefield Suppression in DDS

3.1. *Introduction to DDS scheme*

For the NLC/GLC, several distributions were investigated [51], but a Gaussian distribution received a considerable amount of study. As the inverse Fourier transform of a Gaussian is also a Gaussian distribution, the wakefield decays rapidly according to σ, the prescribed standard deviation of the distribution. However, as each accelerating structure consists of a finite number of cells with a prescribed minimum and maximum iris radius, a sampling interval and a finite bandwidth are imposed upon the Gaussian distribution. In practice, the modes in a given accelerating cavity have a finite frequency separation. This means the wakefield will eventually recohere, according to the minimal separation of these modes. Thus, in practice detuning the modes of a cavity is not expected to be entirely sufficient to ensure adequate Wakefield damping over the complete length of the bunch train. To overcome this inadequacy, and hence prevent the characteristic resurgence in the wakefield, moderate damping is employed. One possible means of affecting this damping is achieved by adding lossy material on each iris to bring down the copper Q of the cavity. Another approach is to couple out a fraction of the dipole radiation through slots from each cell to manifolds attached to the walls of the accelerating structure. The latter was the approach followed by the NLC/GLC and it has some other advantages over other methods of wakefield suppression. In particular, from monitoring the radiation in the manifolds, both the position of the beam can be ascertained and, the alignment of cells can be remotely constructed [52–53]. Furthermore, the damping materials can be located remotely from the cells of the accelerator and, will not be subject to high fields present when breakdown occurs in the cavity. In addition, the pulse temperature rise in the vicinity of the coupling slots, is in general smaller than that in the heavily damped counterpart [54]. A CAD drawing of one of the several accelerating structures that was constructed to test the wakefield suppression properties is illustrated in Fig. 1.

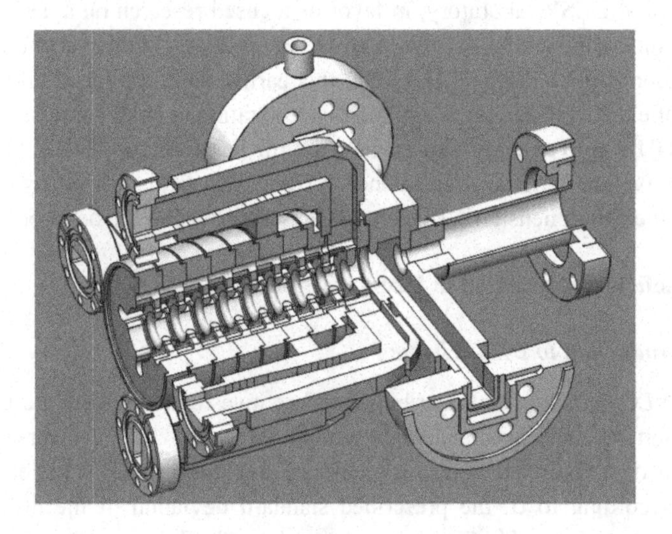

Fig. 1. CAD drawing illustrating a cross-sectional view of the output region of H60VG4SL17A.

This particular structure had a monopole phase advance per cell of $5\pi/6$ and was part of a pair of structures that were measured simultaneously at the ASSET [42] facility at the SLAC National Accelerator Laboratory.

The wakefield excited by a multi-bunch particle beam can be suppressed by ensuring the modes excited in each of the cells of the structure differs in a prescribed manner. There is indeed considerable experience in this method, obtained through more than one and a half decades of research in the area during a collaborative program for the NLC/GLC involving the SLAC National Accelerator Laboratory, KEK National Laboratory, FNAL and LLNL. The mechanism by which the wakefield, excited in multi-cell accelerator structures, couples to the attached manifolds can be understood by considering the characteristic dispersion curves of the cells. This approach is described in the following section.

3.2. *Physical mechanism of mode coupling*

A Brillouin diagram for 3 representative cells out of a 55 cell structure, which operates at a fundamental mode accelerating phase advance of $5\pi/6$, known as H60VG3, is illustrated in Fig. 2. Here the light line is indicated by the black dashed line and, the solid curves represent the dispersion curves for first, middle and last cells prior to coupling to the attached manifolds. The manifold propagates TE waveguide modes which couple to the accelerator via slots in the walls of the cells. Manifold modes are indicated by the dashed curves. The introduction of coupling slots will force a manifold mode and cell mode to be coupled. However, in order to understand the manner which the beam excites modes and couples them to the manifold, the uncoupled curves will be used.

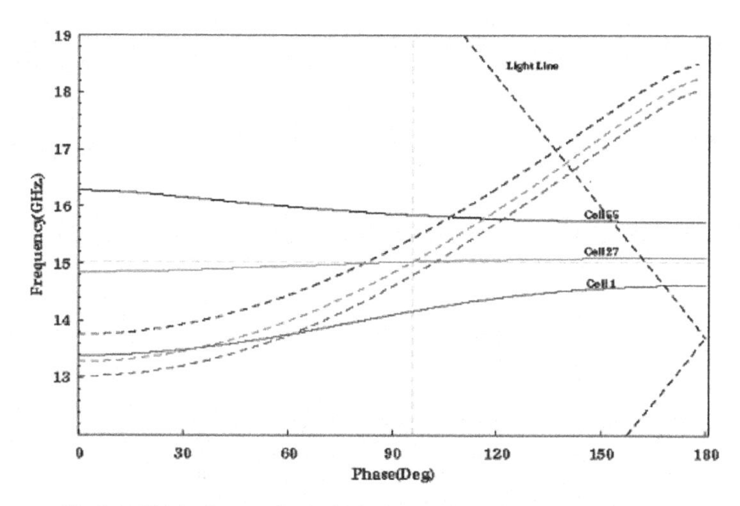

Fig. 2. Brillouin diagram for the high phase advance accelerator H60VG3.

Strong coupling of the dipole modes to the manifold modes occurs where the curves cross. For example, in the centre of the accelerator, at cell 27, the manifold and dipole modes cross at the point indicated by the intersection of the horizontal and vertical green dashed lines (at a frequency of 15.3 GHz and at corresponding phase advance of 95.9 degrees). Interpolating between the dipole curves shown at 0° and 180° indicates that the mode is localized to cells 20 to 34 (outside this domain the modes are in the characteristic "stop-band" region). Also, the horizontal green dashed line intersects the light line at the synchronous point corresponding to cell 20. Thus, the ultra-relativistic beam traversing the accelerator, excites a dipole mode at cell 20, and couples out to the attached manifold, not in the immediate vicinity of the excitation, but further downstream, at the slot located in cell 27. From similar considerations, the coupling to all cells can be understood. The point at which the manifold and dipole modes intersect for each cell is illustrated in Fig. 3, together with the intersection of the light line with the dipole modes, and the location of the 0° and 180° phase advance for each cell. From these curves the mechanism of excitation and coupling can be understood for any of the cells. In general, the dipole mode in a damped and detuned accelerator is excited at a particular cell, is localized (or trapped) to within a limited number of cells, and it subsequently couples out to the attached manifold further down the accelerator.

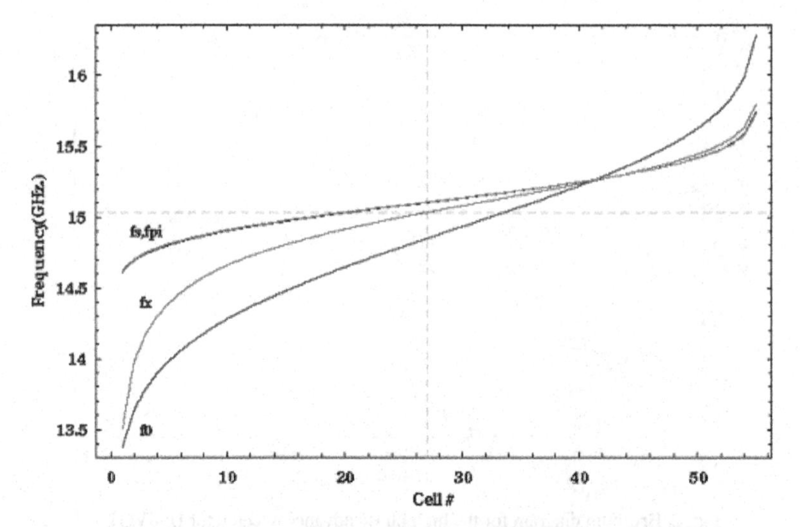

Fig. 3. Mode coupling parameters versus cell number for H60VG3. Here fs indicates the location of the intersection of the light line with the cell mode, fx the intersection of the manifold and cell mode. Also, f0 and fpi indicate the location of the 0° and 180° point in the cells modes.

The wakefield in these damped and detuned accelerating structures can be calculated by various means. In particular, finite element [55] and finite difference [56] computer codes can be used as tools to simulate wakefield excited. However, these codes require substantial resources, in terms of memory and CPU time. In addition, they do not readily lend themselves to rapidly assessing the effect of changes in cell geometry which occur during the course of designing and testing these accelerating structures. Furthermore, during the period the initial structures were being fabricated, no computational tools of this form were available to model the complete accelerating structures. For this reason a circuit model [40, 42] was developed which focused on the dominant first band dipole mode excited in the structures. This was subsequently supplemented with a spectral analysis [42, 58] of the wakefield. This approach is described in a forthcoming section. However, prior to performing a detailed summary of a spectral function calculation of the wakefield based on a circuit model, it is notable that, at short time scales, an approximate method yields reasonably accurate results and this allows a rapid design of wakefield damping for the first few bunches. This approach is described in the following section.

3.3. *Short-time-scale wakefield analysis*

For short times scales, the damping due to the Q resulting from coupling to the manifold can safely be ignored. This is clearly an excellent approximation for the medium damped structures used in the NLC/GLC program (where Q ~500 – 1000) and for those envisaged for the alternate CLIC design. In this region, detuning alone accounts for the decay in the envelope of the wakefield. The envelope of the wakefield is given by:

$$W_t(t) \approx 2\left|\int_{-\infty}^{\infty} K\frac{dn}{df}\exp(i2\pi ft)\,df\right| = 2\left|\int_{-\infty}^{\infty} K(f+\bar{f})\frac{dn}{df}(f+\bar{f})\exp(i2\pi ft)\,df\right| \qquad (1)$$

where K is the characteristic transverse kick factor [16, 41]. Here, a change of variables has yielded a kick factor which is centered about the origin and rapidly oscillating components have been removed to yield the maximum excursion in the wakefield, i.e. the envelope thereof. This approach for the envelope of the wakefield follows a similar analysis developed in [59] for the amplitude of the wake. For example, a uniform distribution,

$$K\frac{dn}{df} = \bar{K}\frac{1}{\Delta f}\Pi\left(\frac{f}{\Delta f}\right), \qquad (2)$$

which is flat over the bandwidth Δf, yields a wakefield:

$$W_t = 2\bar{K}\left|\text{sinc}(\Delta f t)\right|, \tag{3}$$

where sinc(x)=sin(πx)/πx, $\bar{K}$ is the average kick factor, Π (x) = θ (x+1/2) – θ (x – 1/2) and θ is the unit step function (the Heaviside step function [59]). On the other hand, a Gaussian function

$$K\frac{dn}{df} = \bar{K}\frac{1}{\sqrt{2\pi}\sigma}\exp\left(-\frac{f^2}{2\sigma^2}\right), \tag{4}$$

results in a wakefield which decays as,

$$W_t = 2\bar{K}\exp\left[-2(\pi\sigma t)^2\right], \tag{5}$$

The wakefield of additional functional forms may be rapidly computed in this manner. In all cases, the wakefield at the origin (t = 0) is given by twice the area of the associated Kdn/df distribution.

However, this approximate theory does not take into account the finite truncation which inevitably occurs in the chosen Kdn/df distribution. We extend the procedure developed in [59] by taking this truncation into account, for a Gaussian distribution of bandwidth Δf (= $n_\sigma\sigma$). The integral in Eq. (1) performed over finite limits yields:

$$W_t = 2\bar{K}\exp\left[-2(\pi\sigma t)^2\right]\left|\chi[t, n_\sigma]\right|, \tag{6}$$

where the χ factor,

$$\chi[t, n_\sigma] = \frac{\text{Re}\left\{\text{erf}\left[\frac{n_\sigma - i4\pi\sigma t}{2\sqrt{2}}\right]\right\}}{\text{erf}\left[\frac{n_\sigma}{2\sqrt{2}}\right]} \tag{7}$$

arises due to the finite bandwidth of the distribution. As expected, for very short time scales χ is independent of bandwidth, $\lim_{t\to 0}\chi[t, n_\sigma] = 1$. However, when $t \geq n_\sigma/(4\pi\sigma)$, the contribution from truncation becomes significant. It is also clear that as the bandwidth Δf is increased the influence of truncation becomes less serious and it starts to influence only the later points in the bunch train. However, other effects start to play a role later on in the bunch train. In particular, this uncoupled cell model is no longer accurate, as in this case the frequencies of neighbouring cells become coupled. This coupling also affects the kick factors. To include the influence of coupled modes a circuit model description is used. This method is able to include the essential physics of the electromagnetic modes excited and is able to account for rapid remodelling of the wakefield when the structure parameters are varied. In practice, the design of a multi-cell accelerating cavity often requires considerable redesign until the final structure has been obtained. An analysis valid both at short and long time scales is described in the following section.

3.4. *Circuit and spectral wakefield analysis*

A summary of the essential steps in the circuit method to account for coupling in the cell modes is now provided. This circuit model is based on that developed in [59] with a manifold added [42] and is illustrated in Fig. 4 for three representative cells. The method entails obtaining the parameters of a limited number of cells from accurate simulations with finite element or finite difference codes and then interpolating the remaining cells. In general, 8 parameters for each cell are required to characterize the interaction and this is achieved by solving 8 coupled non-linear equations from 8 points on the characteristic dispersion curves. The beam couples to the circuit through transverse magnetic (TM) modes and, as the modes are dipole in character, transverse electric (TE) modes are also excited. These dipole modes are hybrid as they consist of a combination of both TE and TM modes. The hybrid nature of the modes accounts for the necessity of using a two-chain hybrid model; the lower chain representing TM modes and the upper TE modes. Attempting to use a single-chain model, which accounts for TM modes alone, results in a poor fit to points on the Brillouin diagram.

In addition to detuning, the modes are damped through slots cut into the cells and coupled through TE modes to attached manifolds. The modes of the complete structure may then be obtained from the characteristic eigenvalues and eigenvectors of the system which the circuit model describes. However, as the coupling is large and, the frequency shift due to fabrication errors is also potentially large, an eigenvalue approach is not sufficient. For this reason we use a spectral function method [42, 58], which essentially consists in forming the impedance of the cavity, and then taking the Fourier transform of this impedance to obtain the wakefield. This method is able to accommodate large frequency shifts, Ohmic losses and strong mode coupling. A summary of the method to obtain the impedance and wakefield is now provided which begins with the eigensystem.

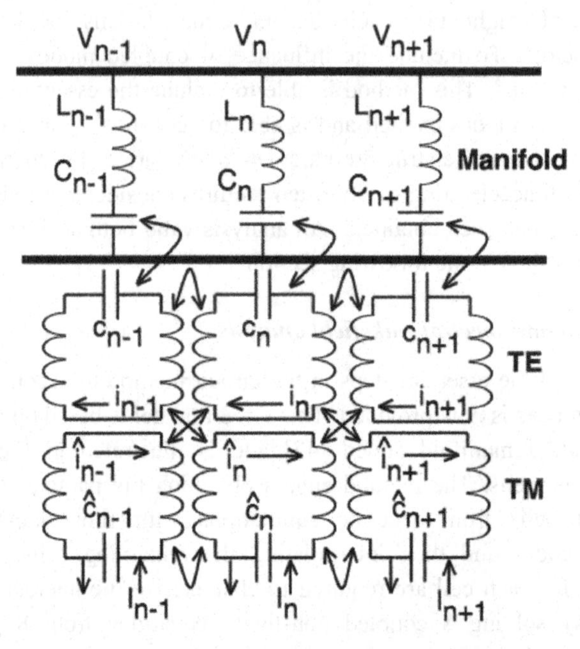

Fig. 4. Circuit model of dipole wakefield coupled to a waveguide-like manifold (obtained with finite element simulations to dispersion curves corresponding to the circuit model).

The equation describing the eigensystem, obtained from the circuit model applied to all N cells is given below:

$$\begin{pmatrix} \hat{H}-\lambda U & H_x^t & 0 \\ H_x & H-\lambda U & -G \\ 0 & G & -R \end{pmatrix} \begin{pmatrix} \hat{a} \\ a \\ A \end{pmatrix} = \begin{pmatrix} \lambda B \\ 0 \\ 0 \end{pmatrix} \tag{8}$$

where $f = \omega/2\pi = 1/\sqrt{\lambda}$ refers to frequency, H ($\hat{H}$) describes TE (TM) modes, H_x^t is transpose of the cross-coupling matrix (between TE-TM modes), G and R are manifold coupling matrices and U is an identity matrix. All these individual elements are themselves matrices of dimension N x N. The TE (TM) eigenvectors in the accelerator cavity are given by a ($\hat{a}$) and the TE modes in the manifold by the vector A. The beam excitation is described by the B vector. The eigenvalues of the matrix in Eq. (8) are complex and are related to the mode frequencies by $f = \omega/2\pi = 1/\sqrt{\lambda}$. Also, as the matrix elements G and R are dependent on frequency, the eigensystem is non-linear and must be solved numerically through an iterative scheme. The real component of each frequency corresponds to the mode frequency. The Q obtained from coupling the accelerator cavity to the waveguide-like manifold is obtained from half the ratio of the mode frequency to the imaginary frequency [61]. The matrix has 3N eigenvalues, corresponding to the TE and TM modes of the cavity and to those modes in the manifold.

A perturbation solution for the complex frequencies can be obtained and is discussed in Appendix C. This will be expected to be reasonably accurate, provided the shift in frequencies is small compared to uncoupled spacing of frequencies. This shift in the spacing of cell frequencies is due to the inclusion of inter-cell coupling and, coupling to the manifold through slots. An example of this procedure applied to the structure DDS1, is illustrated in Fig. 5, where the perturbation approach is compared with solving the eigensystem iteratively. The frequency shift ranges from ~ 2 MHz to ~ 9 MHz and, as the mode spacing is 7 MHz, the original assumption of a small perturbation is invalidated. The corresponding Q values range from 400 to 1400 and can not be considered a small shift about the perturbation solution (which ranges from 600 to 1100). Thus, in order to properly determine the frequencies and Q values, the eigensystem must be solved itertively. However, the iterative method of solving for each eigenvalue is a somewhat cumbersome process, as there are 3xN modes (618 for DDS1) and due to strong coupling, the modes are also shifted as they are no longer located after each other in sequence. For this reason a spectral approach was developed to enable a rapid determination of the modes. This spectral approach essentially consists in forming the impedance of the cavity and then taking the inverse Fourier transform to obtain the wakefield. Wall losses are included in the analysis by providing a resistor in series with each of the L-C circuits and also in the transmission line which represents the manifold. This adds an imaginary term to the diagonals of the corresponding H-matrices, but it has the benefit of broadening the associated resonant peaks in the spectral function and thus reduces the required sampling of frequency points. To condense notation this is written in the form:

$$\left(\bar{H} - \lambda U\right)\bar{a} = \lambda \bar{B} \tag{9}$$

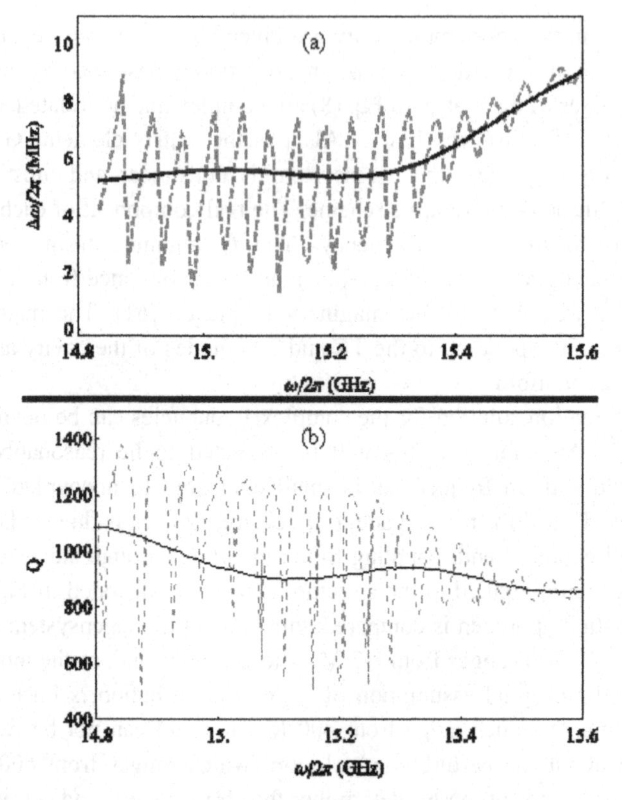

Fig. 5. Frequency shift (a) and Q (b) versus mode frequency calculated from a perturbation method (solid black line) and from an iterative method (red dashed line) for DDS1.

The three elements of Eq. (9) are given by:

$$\bar{H} = \begin{pmatrix} \hat{H} & H_x^t & 0 \\ H_x & H & -G \\ 0 & G & -R \end{pmatrix}, \ \bar{a} = \begin{pmatrix} \hat{a} \\ a \\ A \end{pmatrix} \ \text{and} \ \bar{B} = \begin{pmatrix} B \\ 0 \\ -A \end{pmatrix} \tag{10}$$

The impedance is obtained from [57]:

$$Z(\omega) = \frac{1}{2\pi^2} \sum_{n,m}^{N} \sqrt{K_s^n K_s^m \omega_s^n \omega_s^m} \, \exp[(j\omega L/c)(n-m)] \tilde{H}_{nm} \tag{11}$$

where the 3N by 3N matrix $\tilde{H}$ is given by :

$$\tilde{H} = \bar{H}(U - \lambda^{-1}\bar{H})^{-1} \tag{12}$$

and L refers to the length of a single accelerating cell, and K_s^n is the single-cell transverse kick factor evaluated at the synchronous frequency $\omega_s^n/2\pi$ for mode n. The transverse wakefield (i.e. wake potential per unit length) for a particle trailing a distance s behind a velocity c drive bunch (per unit drive bunch charge per unit drive bunch displacement) may be written

$$W_{\perp}(s) = \frac{1}{2\pi} \int [Z(\omega - j\varepsilon)\exp[(js/c)(\omega - j\varepsilon)]d\omega \tag{13}$$

Here ε refers to a small displacement from the real axis and is introduced to broaden the resonant peaks. In the presence of suitable damping, this offset is not required and in practice the introduction of Ohmic damping is able to broaden the resonances sufficiently to allow the impedance to be adequately sampled. Finally, it should be noted that the circuit model introduces a small non-physical precursor signal and in order to minimize this effect a causal wakefield is introduced. This consists in subtracting the wake ahead of the bunch, which from causality considerations should be zero [42]. The causal wakefield is defined as:

$$W_c(s) = H(s)[W_{\perp}(s) - W_{\perp}(-s)] \tag{14}$$

W_c equals $W_\perp$ for s > NL and vanishes for negative s. W_c is a more faithful representation of the wakefield in the accelerating structure than the strict equivalent circuit model. From causality, for an ultra-relativistic beam there cannot be a wakefield ahead of the exciting bunch and therefore $W_\perp(-s)$ for s>0 is identically zero:

$$W_{\perp}(-s) = \frac{1}{2\pi} \int_{-\infty}^{\infty} Z(\omega + j\varepsilon)\exp[(js/c)(\omega + j\varepsilon)]d\omega \tag{15}$$

which leads to

$$W_{\perp}(s) - W_{\perp}(-s) = \frac{j}{\pi} \int_{-\infty}^{\infty} \mathrm{Im}\{Z(\omega - j\varepsilon)\}\exp(j\omega s/c)\,d\omega \tag{16}$$

$$= \frac{2}{\pi} \int_{0}^{\infty} \mathrm{Im}\{Z(\omega + j\varepsilon)\}\sin(j\omega s/c)\,d\omega \tag{17}$$

To include the contribution of poles on the real axis (with real residue) in (16) and (17) we interpret:

$$\mathrm{Im}\{(\omega \pm j\varepsilon - \omega_0)^{-1}\} = \mp\pi\delta(\omega - \omega_0) \tag{18}$$

and define 4Im{Z(ω+jε)} as the spectral function S(ω) of the wake function. Thus we have

$$W_c(s) = \frac{\theta(s)}{2\pi}\int_0^\infty S(\omega)\sin(\omega s/c)\,d\omega \tag{19}$$

We note further that the wake envelope function $\hat{w}(s)$ associated with W_c, representative of the maximum excursion in the wakefield, is given by

$$\hat{W}(s) = \frac{\theta(s)}{2\pi}\left|\int_0^\infty S(\omega)\exp(\omega s/c)\,d\omega\right| \tag{20}$$

The envelope provides a measure of the worst case wakefield and this proves to be a helpful criterion in designing accelerating structures.

It is interesting to note that the undamped case, corresponds to a sum over discrete eigenmodes, and is obtained by setting the coupling matrix (G in Eq. 10) to zero, Z is real on the real axis and contains a set of poles on the real axis at the modal frequencies. The spectral function is then a sum of delta functions:

$$S(f) = 2\sum_p K_p\delta(f - f_p) = 2K_n dn/df \tag{21}$$

These techniques were employed to design a DDS series starting in 1996 with DDS1 and culminating in 2004, in a pair of structures known as H60VG4SL17A(B).

3.5. *Conspectus of DDS NLC/GLC wakefield suppression*

The initial work in this area entailed detuning each of the modes in a 2 meter long structure, consisting of 206 cells, such that the iris radius followed an erf distribution with cell number. In this manner the iris radius decreases smoothly down the accelerator. The resulting wakefield will initially decay with an approximately Gaussian distribution. However, this is not expected to be an ideal representation of a Gaussian for several reasons. In particular, the endpoints of the Gaussian, defined by the end cells in the structure, will give rise to a truncation in the Gaussian function. Furthermore, as there a finite number of cells which sample the function then the modes which constitute the wakefield will eventually recohere. This recoherance in the modes will result in

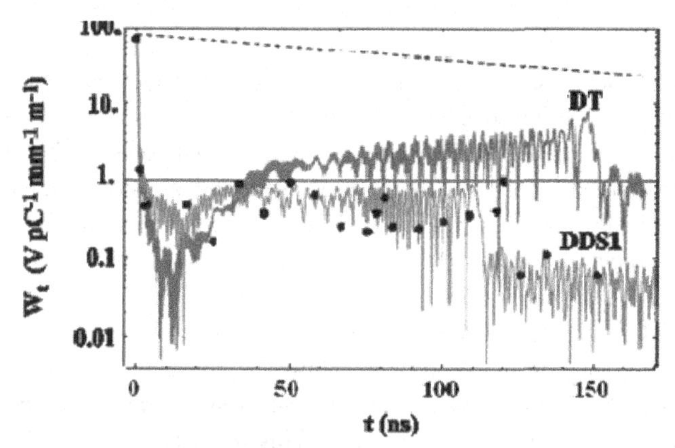

Fig. 6. Simulation of W_t, envelope of wakefield, for DDS1 (in red), the first damped and detuned structure, compared to a detuned structure DT (in blue). Also indicated are points obtained from an experimental measurement of the wakefield and, the damping (indicated by the black dashed line) which is attributable to purely Ohmic losses.

a resurgence in the wakefield. The first measurement on DDS1, an accelerating structure provided with both detuning in the cell frequencies and manifold damping, is illustrated in Fig. 6 along with theoretical predictions based on the circuit model and a spectral analysis of the wakefield. The cells were designed such that the mode density function dn/df, rather than Kdn/df, was prescribed to be Gaussian (details of this method are described in Appendix A). In this case, the initial decay is indeed ~ Gaussian. However, finite sampling gives rise to a resurgence in the wakefield and this is evident for both DDS1 and for the DT version of the accelerator in the region of ~11 m. The manifold damped structure, DDS1, also suffers from an additional feature which dilutes wakefield suppression further due to reflections which occur at the microwave ceramic window to the higher order mode (HOM) loads. These windows had a reflection coefficient $S_{11} \approx 0.1$. Even though this is equivalent to a very small amount of power being reflected (99% transmitted) the wakefield is nonetheless sensitive to S_{11} and hence the suppression in the wakefield is compromised. Simulations indicate that well-matched HOMs would enable the wakefield 5 m behind the first bunch to be at a level of ~0.2 V/pC/mm/m, rather than the predicted and observed level of ~1 V/pC/mm/m. For the NLC/GLC, beam dynamics simulations indicated that the wakefield must be below 1 V/pC/mm/m at the location of each bunch in the train, in order that the emittance dilution be kept below 5%. Thus, DDS1 has met these limits, but manufacturing more than sixteen thousand, as required for a linear collider with a centre of mass energy of 500 GeV, may introduce frequency errors which couldcause the wakefield to fall short of this criterion.

The next major set of structures, DDS3 and DDS4, dispensed with these windows in order to improve overall suppression of the wakefield. Indeed, all subsequent structures were also designed without windows in the HOM load region. Furthermore, the kick factor [16] weighted density function Kdn/df, rather the density function, was prescribed to be a Gaussian function. This enables the spectral function (the inverse transform of the wakefield) to be symmetrical and the minima in the wakefield to be well-defined. The bandwidth and standard deviation of the Gaussian was also modified with a view to optimising the decay in the wakefield. However, in manufacturing each of the 206 cells, which composed DDS3 and DDS4, a significant random error was introduced during the machining process. This impacted the overall wakefield suppression and prevented an observation of the well-defined minima in the wakefield. Even though it is far from the design ideal, it still is below 1 V/pC/mm/m and it is illustrated in Fig. 7. There errors obtained during the fabrication of the cells were not well defined and so errors were introduced at a nominal level into the spectral function model of the structure and increased until the wakefield of the model matches the measured values. This led to ascribing an rms error of 12 MHz to the cell frequencies. However, as the minimum mode spacing is ~ 8MHz, the wakefield is adversely affected due these large random errors introduced during the manufacturing process. The minimum mode spacing effectively sets the maximum allowable frequency errors. Furthermore, the monopole accelerating mode has stringent requirements on acceptable frequency errors which are imposed in order to obtain an acceptable efficiency in accelerating the charged particle beam. For the NLC/GLC this imposed an rms frequency error of no more than ~ 1 MHz on the monopole mode. Once this limit is imposed on the manufacturing process, similar frequency bounds will follow for the dipole mode.

The next series of structure focussed on improving the shunt impedance and Q factor of the accelerating mode. This was achieved by carefully shaping the iris and cavity of each cell of the accelerating structure. In this manner, the shunt impedance was improved by ~10%. This 206 cell accelerator was known as RDDS1. However, in fabricating this structure systematic errors, rather than

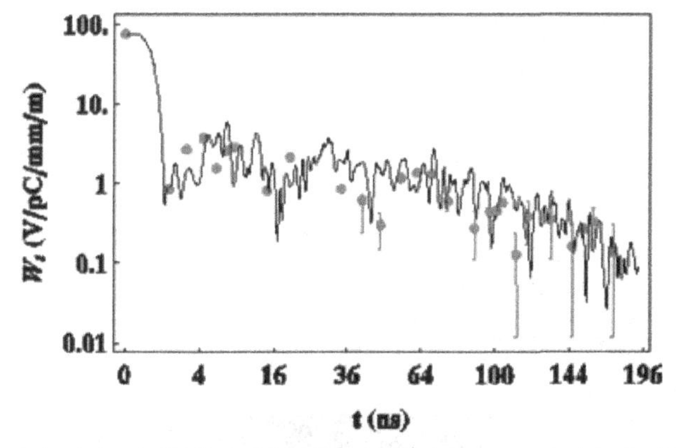

Fig. 7. Spectral simulation (solid line) and experimental measurements (red dots) of the envelope of the wakefield for DDS3. Random errors in the cell frequencies are also included in the simulation with an rms value of 12 MHz.

random errors, were inadvertently introduced into a limited number of cells in the structure. These errors occurred during the brazing process in which the cells are heated to a temperature a little over 1000°C. During this process the structure is supported with a series of stainless steel collars. These collars had a different expansion coefficient from the copper structure and hence they caused a larger expansion of the cells in this region. This expansion was confirmed, prior to measuring the wakefield in ASSET [27], by using a Coordinate Measurement Machine (CMM) to record the outer radius; the resulting inner radius deviation was inferred from that of the outer radius. This gave a predicted shifting in the synchronous frequencies of the middle cells of approximately 30 MHz. This was subsequently confirmed with a "bead-pull" phase measurement of the structure [62]. This systematic shift in the cell frequencies is particularly damaging as it is located in the centre of the structure, theregion of minimum mode spacing, which is ~7 MHz. The mode spacing together with the kick factor weighted density of modes, Kdn/df, are illustrated in Fig. 8. Frequency shifts of a similar magnitude also occurred in the end cells. However, as the mode spacing towards the ends of the distribution is more than 100 MHz the frequency errors had little impact on the wakefield. The resulting experimental measurement of the wakefield together with that of the predicted wake for DS and RDDS1 is illustrated in Fig. 9. The circuit model provides a good prediction of the wakefield as the general properties of the wakefield are reproduced. Replacing the supports with materials of similar expansion coefficients to that of copper will remove this frequency shift and this was incorporated in the next series of structures.

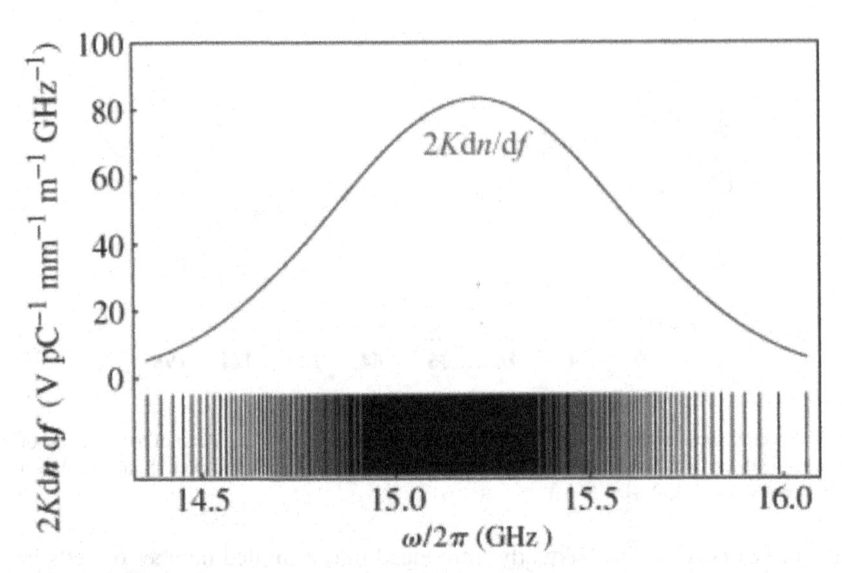

Fig. 8. Twice the smoothed undamped kick factor weighted density function, 2Kdn/df, (in blue) together with the location of individual modes (indicated by vertical black lines).

The next series of structures were designed with a view to achieving high gradient operation, whilst minimising electrical breakdown, and were known as the high phase advance or HDDS series. Inspection of the (R)DDS series indicated that severe damage due to electrical breakdown had been occurring in a number of cells and this motivated the HDDS accelerators. The initial design was prompted from a realisation that the worst damage occurred in the cells with high group velocity and hence the group velocity of subsequent structures was reduced by moving from a $2\pi/3$ phase advance per cell to a $5\pi/6$ phase advance. Furthermore, the input power was reduced by limiting the number cells per accelerating structure to 55, rather than the 206 used in all previous structures. In addition, the ratio of surface field to accelerating monopole field was carefully optimised to be ~2 over the complete accelerating structure by modifying the cell geometry. This limits the high electric field region along the surface of the cavity. Other issues affecting electrical breakdown include surface processing, and gradually bringing the accelerating field up to the required value over a period of time by appropriate high field processing. A detailed discussion of these issues is beyond the scope of this paper. However, breakdown issues in the context of high gradient accelerator cavities are a topic of ongoing research and we refer the reader to [63].

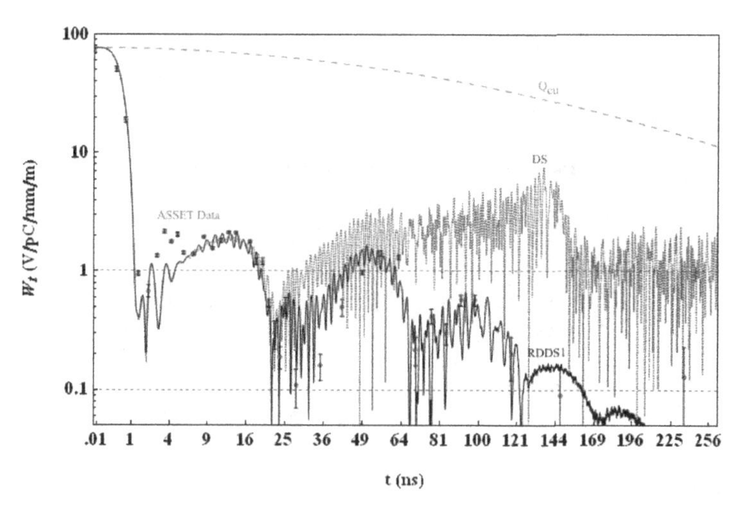

Fig. 9. Envelope of transverse wakefield due to frequency errors obtained in fabricating a complete 206 cell structure. The predicted wakefield is indicated by the solid black line and the red dots correspond to experimentally determined values. The design goal of 1 V/pC/mm/m and the limit in the accuracy of the measurement of 0.1 V/pC/mm/m are indicated by dashed lines.

The wakefield in reduced-length accelerating structures, with a concomitant reduction in surface field, were known as H60VG4SL17A/B. These structures were 60 cm in length and consisted of 55 cells. However, as these structures had an almost 4-fold reduction in the number of cells compared to DDS1, the sampling of the uncoupled kick factor weighted density function, Kdn/df is expected to be severely limited. Indeed, this limited sampling dilutes the effectiveness of the wakefield suppression significantly. For this reason, modes from successive accelerating structures were interleaved. Four-fold interleaving allowed excellent wakefield suppression to be achieved as this allowed the Kdn/df function to sample mode than 220 cells. In order to verify the efficacy of the design, a two-fold interleaved design was performed. This design initial was made by choosing the uncoupled cell frequencies, according to the coupled mode frequencies. As discussed earlier, the coupled cell frequencies determine the long-term behaviour of the wakefield and, the uncoupled ones, the short term behaviour over the region of the first few bunches in the train. The two structures, tested simultaneously at ASSET, were part of a non-fully optimised design which relied on the frequencies from the uncoupled optimisation. The results from both the prediction, based on the circuit function analysis, and experimental values, are displayed in Fig. 10. It is clear that the circuit model provides a good description of the wakefield excited by the beam. Furthermore, this wakefield, even though non-optimum, satisfies the beam dynamics criterion which, for the NLC/GLC prescribed that the wakefield must be below 1 V/(pC.mm.m) at the location of each bunch.

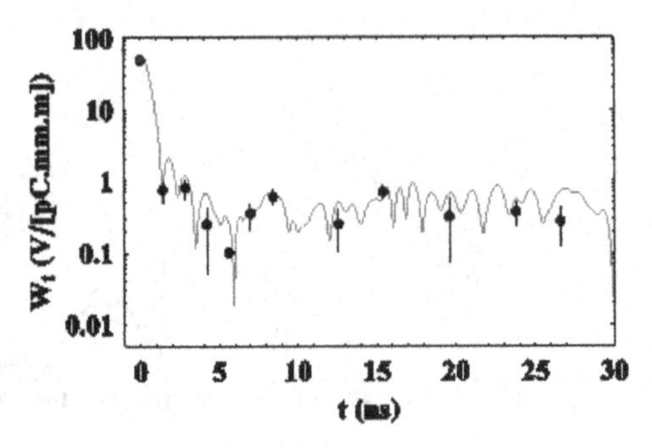

Fig. 10. Envelope of wakefield for H60VG4SL17A/B. The wake predicted with the circuit model is indicated with the solid red line and the experimentally determined values by points.

Non-interleaving does not allow adequate suppression, as in this case after several trailing bunches the wakefield rises above unity. This was the final accelerator structure, designed and tested as part of the NLC/GLC program, equipped with both manifold wakefield suppression and detuning in the cell frequencies. After this structure was tested, a decision was made by the International Technical Review Panel (ITRP) [64] to shift the focus from a dual effort on SC and NC research towards a single concerted approach based on SC technology with an anticipated centre of mass energy reach a the collision point of 0.5 TeV with a potential upgrade to 1 TeV.

However, in order to reach high energies, a realistic path to this goal is provided by higher gradient linacs. The CLIC project aims at a centre of mass collision of 3 TeV using a loaded accelerating gradient of 100 MV/m. The present baseline design for wakefield suppression in the CLIC linacs is to heavily damp the higher modes excited in the cavities [27]. An alternate design, based on moderate wakefield suppression (Q ~ 500) together with strong detuning is being developed. Here the challenge is to achieve adequate wakefield suppression at the first and subsequent trailing bunches, with a more stringent inter-bunch spacing of 0.5 ns (rather than the 1.4 ns used in the NLC/GLC). Research on a moderately damped CLIC structure has reached the stage of a full rf and mechanical design has been completed which satisfies both the wakefield and e.m. field breakdown constraints. The work in this area is discussed in the following section.

3.6. *DDS wakefield suppression applied to the CLIC linacs*

Prior to discussing the geometry utilised in the final design for the upcoming test structure, the initial results are discussed. Two main preliminary structures were investigated. The initial design focussed on satisfying the wakefield suppression demands and this necessitated ascribing a bandwidth of 3 GHz to the first dipole band. A design incorporating 25 cells per structure, utilising 8-fold interleaving invoking a DDS methodology resulted in a well-suppressed wakefield, and is illustrated in Fig. 11.

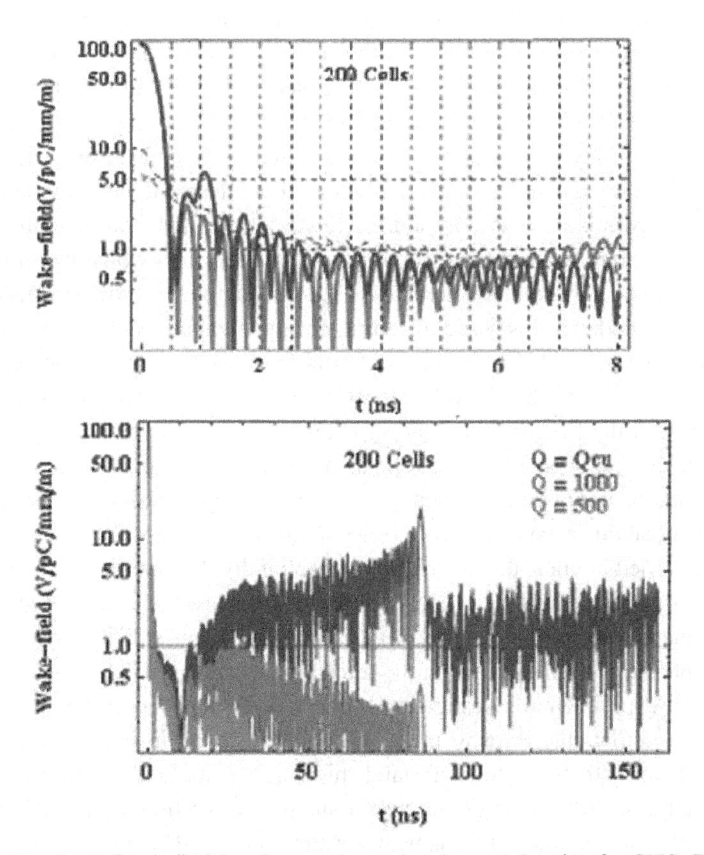

Fig. 11. Envelope of wakefield based on moderate damping and detuning for CLIC. Eight-fold interleaving of a 25-cell structure results in 200 equivalent cells. The wakefield with $Q \sim \infty$ for the first 16 bunches is shown in (a) using an uncoupled model (in blue) and a coupled model (in red). Also shown in (b) is the wake for all subsequent bunches with various Q values.

However, this dipole bandwidth necessitated using a range of cavity geometrical parameters which gave rise to unacceptably large surface electric field on the walls of the cavity (together with a large group velocity in the monopole mode). Thus, even though this structure meets the beam dynamics requirements imposed on wakefield suppression, it fails to fullfill the electrical breakdown criterion [65] which limits the electric and magnetic field surface field. This electrical design constraint limits the electrical breakdown to acceptable levels and ensures the pulse temperature rise on the surface of the cavity is minimised. The present design limits the ratio of the surface electric field to the accelerating field to be no more than approximately two and the pulse temperature rise to 56°K. A further design explored maintaining similar end-cell iris diameters as that in the CLIC baseline design, but with the proviso that the bunches are located at the locations of the zero crossing in the wakefield [27]. The feasibility of this design is closely tied to manufacturing tolerances. A detailed beam dynamics study including the effects of both random and systematic errors is necessary to address this issues. Initial studies suggest this method is a potential solution to suppressing the wakefield whilst the effect on beam emittance dilution is tolerable. A more complete study is needed to categorically address this issue. A third design entailed both modifying the bandwidth to 3.6σ (or $\sim 13.7\%$ of the central frequency) and the bunch spacing from 6 rf cycles to 8 rf cycles (~0.67 ns). Here we adopted eight-fold interleaving of the frequencies of the structures. The beam dynamics constraints, predicated by the short-range wakefield, necessitated a concomitant reduction on the particle population in each bunch. This design also is mindful of the electrical breakdown constraints. In order to ensure the surface electric field is within specified bounds, the outer walls of each individual cell is modified from circular to elliptical [66]. The final wakefield, compared with that of structure without manifolds, is illustrated in Fig. 12. It is evident that the wakefield is well-suppressed for all bunches in the train. As in the NLC/GLC series of DDS designs, an experimental investigation of the wakefields is necessary in order to verify the simulations. Prior to these wakefield tests, an investigation into the ability of the structure to withstand high gradients and high powers is anticipated at CERN. For this purpose a single accelerating structure is in the process of being built. A mechanical design, illustrated in Fig. 13, has been completed. This exploratory structure utilises mode launcher couplers. Additional structures will need to be fabricated to provide the necessary statistics on the influence of manufacturing tolerances on breakdown and also to investigate the wakefield suppression properties with interleaving of successive structures.

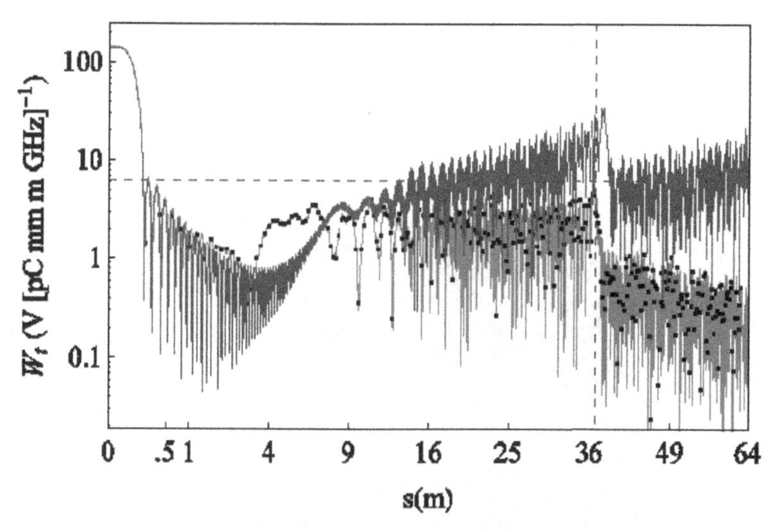

Fig. 12. Envelope of wakefield based on moderate damping and detuning for CLIC. Eight-fold interleaving of a 24-cell structure results in 196 equivalent cells. The envelope of the wakefield is compared for a detuned structure equipped with manifold damping (CLIC_DDS_E, indicated by the red line) to that without manifold damping (DS, indicated by blue line).

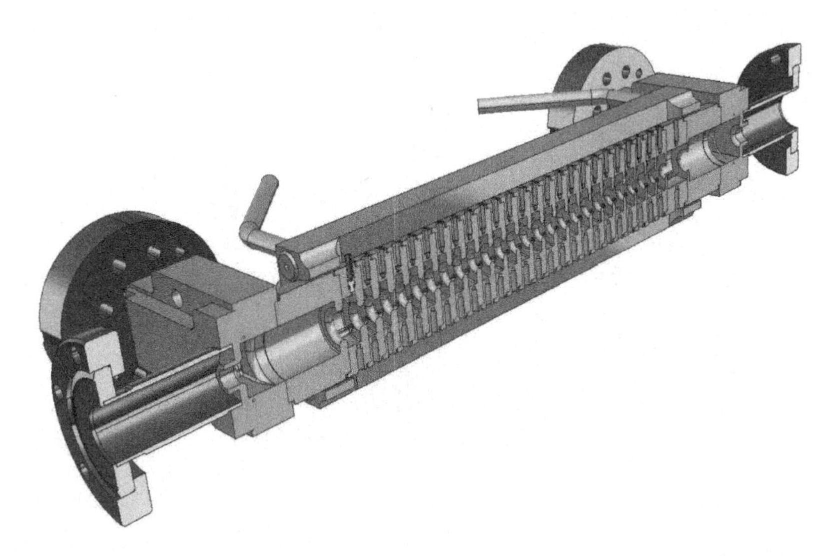

Fig. 13. CAD representation of complete 24 cell accelerating structure CLIC_DDSA.

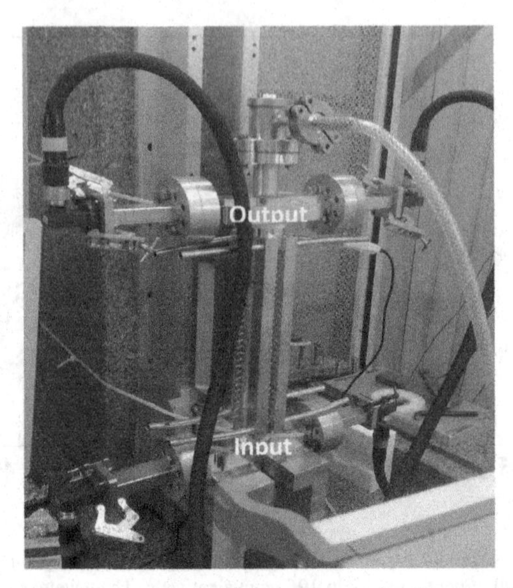

Fig. 14. CLIC_DDSA undergoing tuning. This structure contains 24 main cells together with 2 end cells to facilitate matching into the mode launcher coupler.

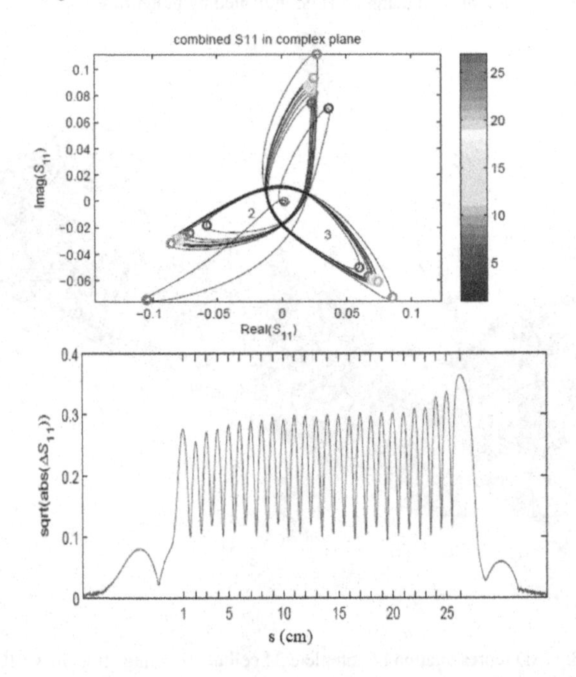

Fig. 15. RF Measurements of reflection coefficient of the final tuned version CLIC_DDSA.

3.7. *HOMs radiated to manifolds in DDS as a structure diagnostic*

A portion of the e.m. field excited by the ultra-relativistic beam is coupled out to the HOM ports and subsequently transported to loads where the radiation is dissipated. By monitoring a small fraction of the dipole component of this energy radiated to the manifolds, both the beam position and cell alignment can be remotely monitored. This is achieved by coaxial loop couplers. The beam can then be centered by moving the beam transverse to the acceleration axis and ensuring the radiation is minimized [69–72]. The dipole component of the power radiated to the manifold is proportional to the square of the offsets. The normalized power, intercepted at the Nth cell, where the HOM coupler is located, is given by:

$$\bar{P} = |A_N|^2 \tag{22}$$

where the variable A_N corresponds to the normalized voltage induced in the manifold at the final cell and is the Nth component of:

$$A = RG^{-1}\left(\lambda^{-1}\bar{H} - U\right)^{-1} X\bar{B} = MX\bar{B} \tag{23}$$

Here X corresponds to the total offset, which includes both the cell offsets (x_i) and beam offset (Δx):

$$X = x_i + \Delta x \tag{24}$$

The normalized power is then given as:

$$\begin{aligned}\bar{P} &= \sum\sum M_i M_j^* (x_i + \Delta x)(x_j + \Delta x) \\ &= \left|\sum M_i\right|^2 \Delta x^2 + \sum\sum\left(M_i M_j^* + M_j M_i^*\right) x_i \Delta x + \left|\sum M_i x_i\right|^2\end{aligned} \tag{25}$$

where each all summations are conducted over all N modes. This power is then minimized at a beam location $\Delta x = \Delta x_0$:

$$\frac{\partial \bar{P}}{\partial \Delta x} = 2\left|\sum M_i\right|^2 \Delta x_0 + \sum\sum\left(M_i M_j^* + M_j M_i^*\right) x_i = 0 \tag{26}$$

The power radiated to the manifold is then given as:

$$\bar{P} = \left|\sum M_i\right|^2 (\Delta x - \Delta x_0)^2 + \left|\sum M_i x_i\right|^2 - \left|\sum M_i\right|^2 \Delta x_0^2 \tag{27}$$

This can be written in the condensed form:

$$\overline{P} = \zeta^2 \left(\Delta x - \Delta x_0\right)^2 + \xi^2 \tag{28}$$

where ζ and ξ are constants to be determined experimentally. This analysis applies for each dipole mode excited within the cavity and hence there will be a series of minima corresponding to each dipole mode component. The influence of the short-range wake on the beam is minimized in this manner by ensuring the HOM power is minimised. The transverse cell-to-cell misalignment can be inferred from the frequency content of HOM signals monitored at the ports. This can be understood by considering the distribution of eigenmodes of which the e.m. field is composed. Most of the eigenmodes are confined to a limited number of cells and hence this offset in the cells will affect a limited number of eigenmodes. In practice this is achieved by moving the beam transverse to the acceleration axis in small steps and recording the power spectrum at each offset. In this manner a complete power spectrum can be built-up and the offset which minimizes each frequency component can be obtained [70–72]. The results of this procedure, applied to DDS1 in which manufacturing assembly errors introduced a ~ 70 μm offset at one end of the structure, are illustrated in Fig. 16. The predictions obtained from the HOM radiation are in good agreement with the mechanical measurement subsequently performed on the structure after it was taken out of the ASSET facility.

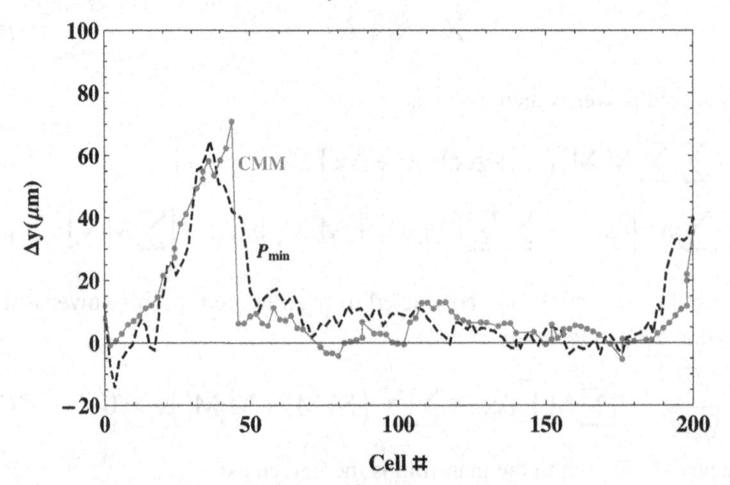

Fig. 16. Cell offsets of DDS1 obtained by coordinate measurement machine (CMM), indicated by red connected dots and, inferred from the energy radiated from the HOM ports (P_{min}), indicated by a black dashed line.

4. Final Remarks

There are two main strategies of wakefield suppression for lepton collider applications. One entails heavy damping, which amounts to designing an accelerator structure which has a small Q values (~10). The other method entails moderate damping, with Qs ranging from 500 to 1000, together with strong detuning in the cell frequencies. The latter, damped and detuned (DDS) method of wakefield suppression, has the potential advantage that, as the wakefield is coupled out through waveguide-like manifolds, it can be dissipated in a location outside the vicinity of the beam and consequently the loads will not be subject to large e.m. fields. Also, in the quest for high gradient acceleration in the CLIC scheme, the DDS approach offers the potential for a lower maximum pulse temperature rise. Furthermore, this approach is relatively mature, in that it has received more than one and half decades of focussed research from several international laboratories for the NLC/GLC programme and, this has resulted in a succession of DDS structures. These structures have been both experimentally tested, with a view to determining their ability to sustain high gradients, and the circuit model of wakefield prediction has been verified with beam-based experiments at the ASSET facility of the SLAC National Accelerator Laboratory.

Beam dynamics simulations conducted on NLC/GLC structures indicates that relatively relaxed random frequency errors, as high as 20 MHz [27], are allowable. This contrasts with the accelerating mode, which has a significantly tighter frequency tolerance of 1 MHz. Thus, provided the fundamental accelerating mode tolerance is met, the higher order mode wakefield tolerance will be relatively straightforward to achieve. Systematic errors are however more restrictive, as this is dominated by the minimal frequency separation – which in general will be a little larger than 5 MHz in the centre of the structure and an order of magnitude larger towards the ends of the structure. A similar suite of beam dynamics simulations, using PLACET [67] and LUCRETIA [68] in particular, is necessary to assess the ability of these CLIC DDS structures to preserve beam emittance under realistic manufacturing conditions.

Finally, it is important to note that, at this stage of the research, both means of wakefield suppression should be pursued in order to assess their applicability to linear colliders. Experimental testing and verification of wakefield predict-tions and ability of the structure to sustain high gradients, using realistic pulse lengths planned for linear collider operation, will be the final indication on the suitability of these techniques.

Acknowledgments

Significant parts of this work were supported by the U.S. Department of Energy under contract DE-AC02-76SF00515. The most recent parts of this research leading to these results on CLIC structures in particular has received funding from the European Commission under the FP7 Research Infrastructures grant agreement no.227579. This document contains material, which is the copyright of certain EuCARD beneficiaries and the European Commission, and may not be reproduced or copied without permission. The information hereindoes only reflect the views of its authors and not those of the European Commission. The European Commission and the EuCARD beneficiaries do not warrant that the information contained herein is capable of use, or that use of the information is free from risk, and they are not responsible for any use that might be made of data appearing herein.

Appendix A: Gaussian Distribution and Mode Separation

In the SBLC study Gaussian detuning was considered as a means of suppressing the long-range wakefield. Also, the first accelerating structure in the damped detuned series known as DDS1, was designed with an uncoupled mode density function proportional to a Gaussian:

$$dn/df \propto \exp\left[-\frac{\left(f-\bar{f}\right)^2}{2\sigma^2}\right] \tag{A.1}$$

where $\bar{f}$ represents the average dipole frequency. Integrating Eq. (A.1) over the complete structure, consisting of N cells, enables the constant of proportionality to be determined:

$$dn/df = \frac{N-1}{\sqrt{2\pi}\sigma \operatorname{erf}\left[\frac{n_\sigma}{2\sqrt{2}}\right]} \exp\left[-\frac{\left(f-\bar{f}\right)^2}{2\sigma^2}\right] \tag{A.2}$$

with n_σ being the frequency bandwidth in units of σ. Modes i and i-1 are described by:

$$\operatorname{erf}\left[\frac{f_i-\bar{f}}{\sqrt{2}\sigma}\right] = \operatorname{erf}\left[\frac{f_{i-1}-\bar{f}}{\sqrt{2}\sigma}\right] + \frac{2}{N-1}\operatorname{erf}\left[\frac{n_\sigma}{2\sqrt{2}}\right] \tag{A.3}$$

Close to the centre of the distribution the modes are spaced by:

$$\Delta f / \bar{f} = \left(f_i - f_{i-1}\right) / \bar{f} = \frac{\sqrt{2\pi}}{N-1}\frac{\sigma}{\bar{f}}\,\mathrm{erf}\left[\frac{n_\sigma}{2\sqrt{2}}\right] \tag{A.4}$$

Here we have used a Maclaurin series expansion: $\mathrm{erf}(x) \sim 2x/\sqrt{\pi}$, valid for $|x| \ll 1$. This enables the location of the recoherence in the wakefield to be determined (~1/Δf). The benefits of interleaving successive structures are evident, as it increases the effective number of cells and ensures the recoherence position will be located further down the bunch train.

Appendix B: Perturbation Analysis of Mode Frequencies

The wakefield is damped by introducing four manifolds, which run collinear to the acceleration axis, to couple out a portion of the e.m. energy radiated by the beam. Setting the coupling to zero, corresponds to closing off the coupling slots, and we refer to this as the unperburbed solution to the eigensystem repesented by Eq. (11). It is convenient to write the matrix description of the coupled modes represented in Eq. (11) as:

$$\begin{pmatrix} H - GR^{-1}G & H_x \\ H_x^t & \hat{H} \end{pmatrix}\begin{pmatrix} a \\ a \end{pmatrix} = \lambda\begin{pmatrix} a \\ a \end{pmatrix} \tag{B.1}$$

Here we have used the relation between the eigenmode in the manifold A and that of the TE eigenmodes in the accelerating structure: $A = R^{-1}Ga$. In order to seek a solution about the unperturbed one, we set the coupling to the manifold to zero, and this corresponds to $G = 0$ (where 0 is in fact an $N \times N$ matrix of zeros) to give the unperturbed eigensystem:

$$\begin{pmatrix} H - GR^{-1}G & H_x \\ H_x^t & \hat{H} \end{pmatrix}\begin{pmatrix} a_0 \\ a_0 \end{pmatrix} = \lambda\begin{pmatrix} a_0 \\ a_0 \end{pmatrix} \tag{B.2}$$

Each element of the above 2×2 matrix is dimensioned $N \times N$. Thus, the zero order solution to Eq. (B.2) gives $2N \times 2N$ eigenvalues and eigenvectors. Provided the system is loss-less Eq. (B.2) is a linear eigensystem and is solved by standard methods. However, Ohmic losses make the matrix in Eq. (B.2) a function of frequency $(=\lambda^{-1/2})$ and hence the solution of a non-linear eigensystem is required. In practice, for X-band structures for example, the Ohmic Q ~ 6500 for a dipole center frequency of ~ 15 GHz and in this case the non-linear eigensystem may be solved iteratively.

The first order perturbation solution to the eigenvalues of Eq. (B.1) is given by:

$$\frac{\delta\omega}{\omega_0} = \frac{\omega_0^2}{8\pi^2}\frac{\bar{a}_0^\dagger G R^{-1} G \bar{a}_0}{\bar{a}_0^\dagger \bar{a}_0} \tag{B.3}$$

where the frequency has been expanded as $f=(\omega_0+\delta\omega)/2\pi$, $\bar{a}_0 = \begin{pmatrix} a_0 \\ \hat{a}_0 \end{pmatrix}$ and † refers to the complex conjugate of the transpose of the vector. The real and imaginary components of Eq. (B.3) allow the mode frequency and mode $Q(=\mathrm{Re}[\omega]/2\,\mathrm{Im}[\delta\omega])$ to be readily obtained. The curves in Fig. 5 are obtained in this manner.

Finally, it is worth noting that even though Eq. (B.2) has resulted in a 2N × 2N matrix rather than the original 3N × 3N matrix, when solving for the spectral function it is more convenient to solve for the original 3N × 3N matrix. This is because the 3N × 3N matrix is a sparse matrix and standard techniques may be employed in this case; whereas the 2N × 2N matrix requires an additional matrix inversion (in $GR^{-1}G$) and is computationally more expensive than solving Eq. (9). Thus, for strong coupling it is preferable to solve the original 3N × 3N matrix given by Eq. (9).

References

1. Lyndon Evans and Philip Bryant (eds.), JINST 3 S08001 (2008).
2. Lyndon Evans, New J. Phys. 9 335 (2007).
3. V. Balakin, VLEPP, Proc. LC91, BINP, Protvino, USSR, 1992.
4. P. LeBrun *et al.*, The CLIC programme: towards a staged e+e− linear collider exploring the Terascale, CLIC CDR, CERN-2012-005, 2012.
5. International Linear Collider Technical Review Committee (ILC-TRC) Second Report, 2003, SLAC-R-606.
6. International Linear Collider Technical Review Final Report, August 2004 (http://www.linearcollider.org).
7. R. Brinkman *et al.* (eds.), The TESLA Technical Design Report, March 2001, DESY Report 2001-33.
8. R.L. Geng, Physica C: Superconductivity, 441, 1-2, 145-150, (2006).
9. T. Saeki presented at SMTF collaboration meeting, Fermilab, Batavia, IL, Oct. 5–7, 2005.
10. N. Phinney *et al.* (ed.) 2001 Report on the Next Linear Collider, SLAC-R-571.

11. J. Wang and T. Higo, ICFA Beam Dyn. Newslett. 32:27-46, 2003 (also SLAC-PUB 10370).
12. T. Higo, Progress of X-Band Accelerating Structures, Proc. Of Linac 2010, Tsukuba, Japan, 2010.
13. H. Braun *et al.*, CLIC 2008 Parameters, CLIC-Note 764 (2008).
14. Frank Tecker, Journal of Physics: Conference Series 110, 11205 (2008).
15. Jean-Pierre Delahaye, J. Phys. Conf. Ser. 110:012009, 2008.
16. G. Riddone (private communication).
17. P.B. Wilson, Introduction to wake fields and wake potentials, Contributed to U.S. Particle Accelerator School, Batavia, Ill., Jul 20–Aug 14, 1987. (SLAC-PUB-4547).
18. P.B. Wilson, Rev. Accel. Sci. Tech. 1:7–41 (2008).
19. V. Balakin, S. Novakhatsky, and V. Smirnov, Proc. of 12th Int. Conf. on High Energy Accel., Fermilab, 1983, p. 119.
20. K. Yokoya, Cumulative beam break up in large scale linacs, 1989, DESY Report 86-084.
21. H. Henke, CERN-LEP-RF/88-49, CLIC-Note-78, 1988.
22. A. Grudiev *et al.*, *Proceedings of the European Particle Accelerator Conference (EPAC 06), Edinburgh, Scotland, 2006* (also CERN-AB-2006-028).
23. W. Wuensch, *Proceedings of x-band structures and beam dynamics workshop (XB08), Cockcroft Institute, UK, 1-4 Dec. 2008.*
24. R.M. Jones *et al.*, *Proceedings of the XXI International Accelerator Conference (LINAC02), Gyeongju, Korea, 2002* (also SLAC-PUB-9467).
25. R.M. Jones, *Proceedings of the 22nd Linear Accelerator Conference (LINAC04), Lubeck, Germany, 2004* (also SLAC-PUB 10682).
26. R.M. Jones, *et al.*, New J. Phys. 11 3, 033013, 2009.
27. R.M. Jones, Phys. Rev. ST Accel. Beams 12, 104801, 2009.
28. M. Sands and J. Rees, SLAC Report PEP-95 (1974).
29. F. Caspers, *Bench methods for beam-coupling impedance measurement (Lecture notes in beams: intensity limitations vol 400)* (Berlin, Springer, 1992).
30. N. Baboi *et al.*, *Proceedings of the 8th European Particle Accelerator Conference (EPAC02), Paris, France, 2002* (also SLAC-PUB 9248).
31. R.M. Jones *et al.*, *Proceedings of the 8th European Particle Accelerator Conference (EPAC02), Paris, France, 2002* (also SLAC-PUB 9245).
32. T. Shintake, Japanese J. Appl. Phys. 31:L1567-L1570 (1992).
33. T. Shintake, Sendai Linear Accel. 1992:67-69 (1992).
34. T. Shintake, *Proceedings of 1993 Particle Accelerator Conference, Washington, DC, USA, 1993.*
35. T. Shintake, H. Matsumoto, H. Hayano, *Proceedings of 17th International Linac Conference (LINAC94), Tsukuba, Japan, Aug 21–26, 1994.*
36. H. Matsumoto, T. Shintake, N. Akaska, K. Watanabe, H.S. Lee, O. Takeda, *Proceedings of 5th European Particle Accelerator Conference (EPAC 96), Sitges, Spain, 1996.*

37. H. Matsumoto, T. Shintake, N. Akasaka, *Proceedings of XIX International Linac Conference LINAC98, Chicago, Illinois, USA, 1998.*
38. T. Shintake *et al., Proceedings of the IEEE Particle Accelerator Conference (PAC 99), New York, 1999.*
39. C. Adolphsen, *et al.*, Phys. Rev. Lett. 74:2475–2478 (1995).
40. R.M. Jones *et al., Proceedings of 5th European Particle Accelerator Conference (EPAC 96), Sitges, Spain, 1996* (also SLAC-PUB 7187).
41. C. Adolphsen *et al.*, New York 1999, Particle Accelerator, vol. 5 3477-3479 (also SLAC-PUB-8174).
42. R.M. Jones *et al.*, Phys. Rev. ST Accel. Beams 9, 102001 (2006).
43. M. Dehler *et al., Proceedings of the 7th European Particle Accelerator Conference (EPAC 2000), Vienna, Austria, 2000* (also SLAC-PUB-8941).
44. A. Mosnier, Instabilties in linacs DAPNIA-SEA-93-19, 1993.
45. R. Brinkmann (editor), Linear Collider Conceptual Design Report, Chapter 4: S-Band Linear Collider, 1997 (http://www.desy.de/lc-cdr/).
46. N. Holtkamp, *Proceedings of the 1995 Particle Accelerator Conference (PAC95), Dallas, Texas, 1995.*
47. N.P. Sobenin *et al., Proceedings of 18th International Linear Accelerator Conference (LINAC 96), Geneva, Switzerland, 1996.*
48. P. Hulsmann *et al., Proceedings of 5th European Particle Accelerator Conference (EPAC 96), Sitges, Spain, 1996.*
49. *The Stanford Two-Mile Accelerator*, edited by R.B. Neal (W.A. Benjamin, Inc., New York, 1968).
50. R.H. Helm *et al., Proceedings of Particle Conference, Washington, D.C., 1969* (also SLAC-PUB-563).
51. R.H. Miller (private communication).
52. R.M. Jones, Structure alignment diagnostics from HOM radiation, This Special Issue.
53. R.M. Jones *et al., Proceedings of 20th International Linac Conference (Linac 2000), Monterey, California, 2000* (also SLAC-PUB-8609).
54. Z. Li (private communication).
55. Z. Li *et al.*, X-band linear collider R&D in accelerating structures through advanced computing 9th European Particle Accelerator Conference (EPAC04), Lucerne, Switzerland, 5 - 9 July, 2004 (also SLAC-PUB-10563).
56. W. Bruns, Gdfid L, *Proceedings of Particle Accelerator Conference (PAC97), Vancouver, B.C., Canada, 1997.*
57. R.M. Jones *et al., Proceedings of 5th European Particle Accelerator Conference (EPAC 96), Sitges, Spain, 1996* (also SLAC-PUB-7187).
58. R.M. Jones *et al., Proceedings of 18th International Linear Accelerator Conference (Linac 96), Geneva, Switzerland, 1996* (also SLAC-PUB-7287).
59. K.L.F. Bane and R.L. Gluckstern, Particle Accelerators 42 pp 123-169 (1993).

60. R. Bracewell, *The Fourier Transform & Its Applications, 2nd Edition* (McGraw-Hill Kogakusha, ltd, 1978).
61. J.D. Jackson, *Classical Electrodynamics, 3rd Edition* (John Wiley & Sons, Inc, 1999).
62. J. Lewandowski *et al.*, *Proceedings of 22nd International Linear Accelerator Conference (LINAC 2004), Lubeck, Germany, 2004* (also SLAC-PUB-11193).
63. L. Laurent, High gradient RF breakdown studies, Ph.D. Thesis, University of California, Davis, 2002.
64. International Committee for Future Accelerators (ITRP): http://www.ligo.caltech.edu/skammer/ITRP_Home.htm.
65. A. Grudiev, S. Calatroni, and W. Wuensch, Phys. Rev. ST Accel. Beams 12, 102001, 2009 (XB08 special issue).
66. V.F. Khan, Wakefield and Surface Electromagnetic Field Optimisation of Manifold Damped Accelerating Structures For CLIC, This Special Issue.
67. E. D'Amico *et al.*, *Proceedings of IEEE Particle Accelerator Conference (PAC'2001), Chicago 2001, Illinois, USA, 2001.*
68. P. Tenenbaum, *Proceedings of 2005 Particle Accelerator Conference, Knoxville, Tennessee, 16-20 May 2005* (also SLAC-PUB-11215).
69. M. Seidel *et al.*, Nuclear Instruments and Methods in Physics Research A 404 (1998) 231-236.
70. R.M. Jones *et al.*, *Proceedings of 7th Workshop on Advanced Accelerator Concepts, Lake Tahoe, CA, USA, 1996* (also SLAC-PUB 7388).
71. R.M. Jones *et al.*, *Proceedings of the 1997 Particle Accelerator Conference (PAC97), BC, Canada, 1997* (also SLAC-PUB 7539).
72. S. Döebert *et al.*, *Proceedings of Particle Accelerator Conference (PAC 05), Knoxville, Tennessee, 2005*, (also SLAC-PUB 11206).

Higher Order Mode Damping Simulation and Multipacting Analysis

Liling Xiao

SLAC National Accelerator Laboratory, Menlo Park, CA, USA

When the beam is passing through an accelerator, it will generate higher order modes (HOM), which will affect to the beam quality especially in high energy accelerators such as International Linear accelerator Collider (ILC). In order to preserve the beam quality, HOM couplers are required to be installed to extract HOM power. Most of HOM couplers are 3D complex structures including small features. In addition, many physics process are involved in HOM coupler design such as RF heating and multipacting. Numerical modeling and simulation are essential for HOM coupler design and optimization for successful operation of high energy accelerators. SLAC developed 3D finite element parallel electromagnetics code suite ACE3P can be used to accelerator modeling with higher accuracy in fast turnaround time. In this paper, ACE3P application for HOM damping simulation and multipating analysis is presented for ILC 3.9 GHz crab cavity.

1. Introduction

In high energy or high intensity accelerators, it is very important to minimize the beam induced HOM power and preserve the beam quality through HOM couplers. Based on the damping requirements for different projects, there are many types of HOM couplers developed and used in the past. Some of typical HOM couplers are shown in the Fig. 1. The loop antenna HOM coupler has been adopted by HERA, TESLA, SNS, and CEBAF. It is compact and takes less beamline space. The beam absorber coupler can damp broadband HOM, which is suitable for short bunch applications such as XFEL, ERL, and KEKB-crab cavity. A waveguide coupler has nature filter because of its cutoff frequency. In addition, it can handle high power. Waveguide HOM coupler has been used in CEBAF and ANL SPX deflecting mode cavity.

Usually HOM couplers are complex structures. Effective 3D simulation tool is essential to HOM coupler design and optimization. SLAC developed ACE3P (Advanced Computational Electromagnetics 3D Parallel) is a comprehensive set of conformal, high-order, parallel finite-element electromagnetic codes. Based on high-order curved finite elements for high fidelity modeling, ACE3P can achieve better solution. Implemented on massively parallel computers, ACE3P can solve large complex problems with increased memory at greater speed [1]. ACE3P consists of the six application modules: Omega3P, S3P, T3P, Track3P, TEM3P, and Pick3P that can cover the needs of accelerator design. In the last two decades, ACE3P has been used in a wide range of applications in accelerator science and development including the 3.9 GHz ILC crab cavity design.

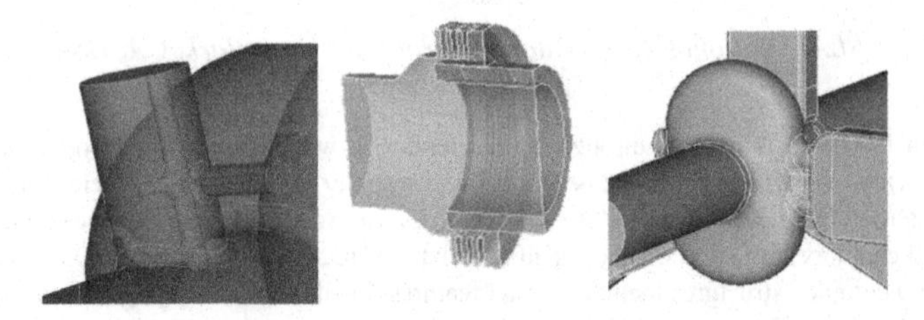

Fig. 1. Three typical HOM couplers — loop antenna coupler (left), beam absorber coupler (middle), and waveguide coupler (right).

The crab cavity design for the ILC beam delivery system (BDS) is based on the 3.9 GHz deflecting mode cavity originally developed at Fermilab [2] for the CKM (Charged Kaons at the Main Injector) beam line as the RF requirements are quite similar for the two machines. Two 9-cell crab cavities operating at 5 MV/m deflecting gradient will be needed for each of the positron and electron beam lines for the ILC [3].

The FNAL 9-cell crab cavity model is shown in Fig. 2. The polarization-flats are formed in the horizontal orientation of the cells (1.5 mm indentation) to split the degeneracy of the horizontal and vertical TM_{110} π-modes by about 9 MHz. The lower frequency polarization mode is in the horizontal plane and chosen to be the operating mode. The bunch is in phase quadrature with the RF so that the head and tail are kicked in the opposite directions by the deflecting mode to realize horizontal rotation (crabbing). To achieve a clean crabbing to the bunches, effective wakefield damping is crucial. In addition to the higher order modes (HOM), the lower order TM_{010} modes (LOM) and the same order vertical TM_{110} π-mode (SOM) also need to be damped through the HOM, LOM, and SOM couplers, respectively.

In this paper, we use ACE3P to simulate and analyse the LOM, HOM and SOM damping results in the Fermilab design. It is found that the LOM, HOM and SOM couplers provide inadequate damping to some modes, and that the sensitivities of the LOM and HOM notch filter gaps are too high that they impose stringent tolerance requirements for tuning. Because of these limitations of the original design, we propose an improved design that can alleviate the aforementioned problems by modifying the LOM and HOM couplers. We will also present multipacting analysis of the couplers to determine if there are any possible high power processing barriers.

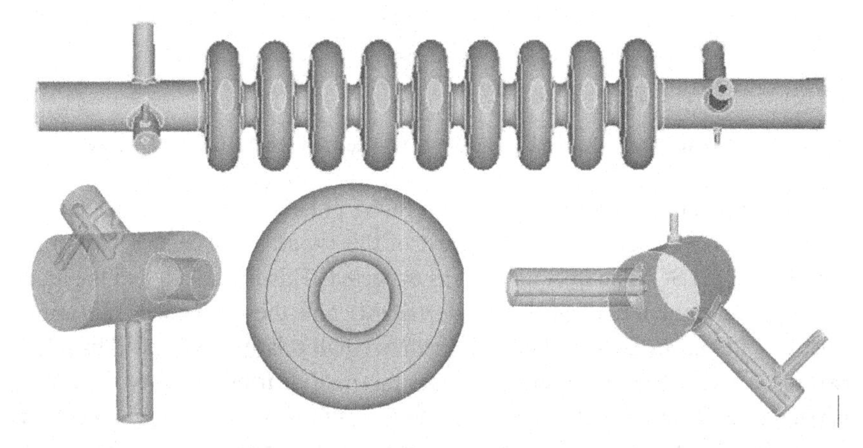

Fig. 2. Model of the FNAL deflecting mode cavity with input and HOM couplers at left end and SOM and LOM couplers at right end.

2. Damping Simulations of the Final Deflecting Mode Cavity

In the original Fermilab design, the fundamental and SOM couplers are situated at opposite ends of the cavity and perpendicular to each other, while the LOM and HOM couplers are oriented at certain angles at opposite ends such that all the modes are efficiently damped. To investigate the effectiveness of this configuration for wakefield damping, we use the complex eigensolver ACE3P-Omega3P to solve for the modes up to the second dipole band, where most of the modes are below the beampipe cutoff frequency of 4.88 GHz. Figure 3 shows the R/Q and Qext of the first monopole and the first two dipole bands. For the monopole band, the modes around 2.83 GHz have relatively high R/Q. For the first dipole band, the R/Q is small except for the operating mode and the SOM mode which also has a high Qext. The second dipole band modes have low R/Q and Qext, and hence their effects on wakefield can be neglected. In general it can be seen that the exiting couplers achieve efficient damping for the LOMs and HOMs. However, it is desirable to improve the damping of other modes such as the SOM.

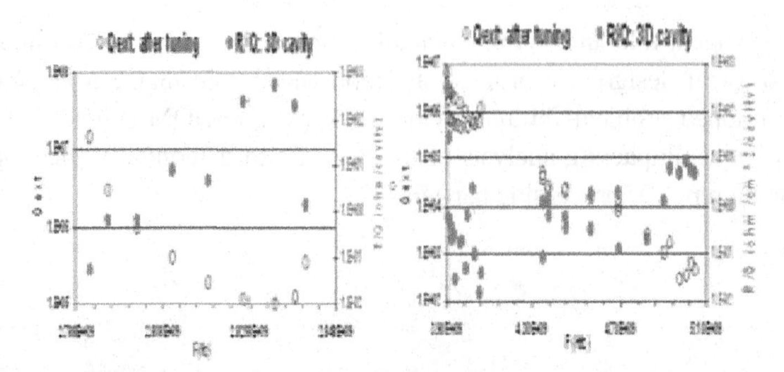

Fig. 3. ACE3P-Omega3P damping results — LOM (left) and HOM (right), respectively.

The critical dangerous mode is the SOM polarized in the vertical plane. Simulations have indicated that the Qext of this mode is not as sensitive to the intrusion of the SOM coupler antenna as expected. This is attributed to the mode coupling between the vertical π-mode (SOM) and the horizontal 7π/9 mode since the frequency spacing between them is small relative to the widths of the two resonances. The mode mixing causes the field distribution to twist and the maximum electric field of the SOM no longer aligns with the SOM coupler as shown in Fig. 4, reducing the effectiveness of the damping. In addition, the mode mixing may cause x-y wakefield coupling which is presently being investigated. This problem however can be resolved by modifying the cell shape to decouple these modes [4].

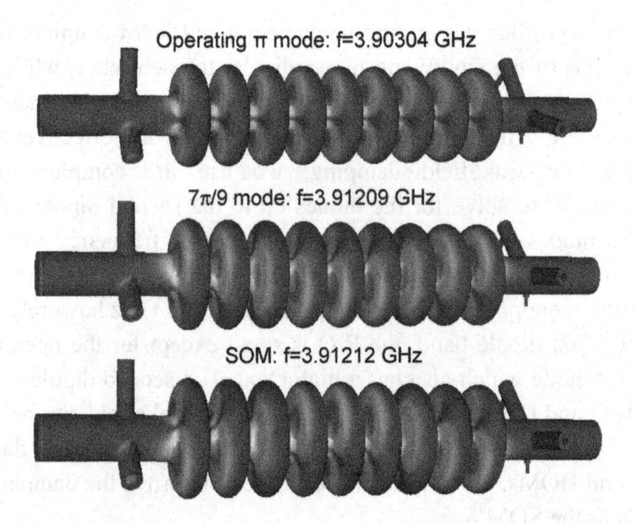

Fig. 4. Mode patterns of the TM110-π dipole pair and the horizontal TM110-7π/9 mode.

In practice, the notch filters in the LOM and HOM couplers need to be tuned to the fundamental mode frequency for the rejection of input power from the fundamental coupler. Using the S-parameter code ACE3P-S3P to determine the transmission coefficients for different notch gaps, the tuning characteristics of the LOM and HOM notch filters are plotted in Fig. 5. The sensitivities of the LOM and HOM notch filters are found to be 2.2 MHz/μm and 1.6 MHz/μm, respectively, an order of magnitude higher than that of the TESLA TTF cavity. This may affect the rejection of the fundamental mode power and cause damage to the LOM and HOM couplers.

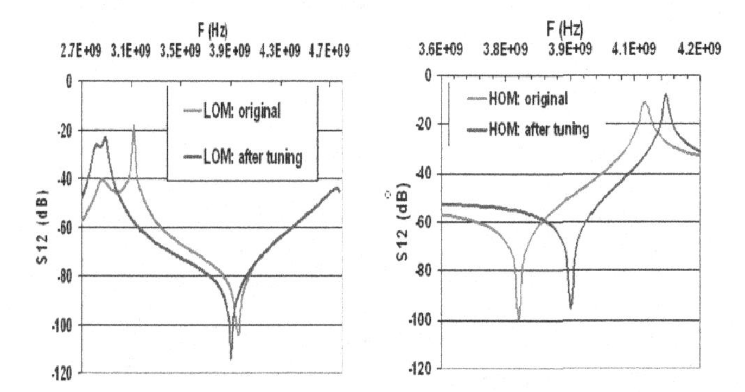

Fig. 5. ACE3P-S3P results on tuning characteristics of the notch filters in the LOM (left) and HOM (right) couplers. The red curve is for the original design and the blue curve is after adjustment of the notch gaps of the LOM/HOM couplers and the hook length of LOM coupler.

3. LOM and HOM Coupler Optimizations for the ILC Crab Cavity

The existing LOM coupler is oriented 135° in the azimuthal direction from the input coupler. It will couple out both the operating TM_{110} π-mode and the TM_{010} monopole modes, and hence the notch filter is required to reject the operating mode. By placing the LOM coupler to the opposite side of the SOM coupler in the vertical plane, the coupling of the operating mode will vanish while those of the monopole modes will not be affected because of their azimuthal field symmetry. This arrangement allows for the elimination of the notch filter and the pickup coax, and hence simplifies the design. The coupler can be adapted to a larger output coax at the end which is advantageous for power handling.

Without a notch filter in rejecting the fundamental mode, both the LOM and SOM coupler adjustment for misalignment need to be investigated. Simulation showed that when the central conductor of the new LOM coupler shifts by 1 mm, the operating mode's Qext can still reach the order of 10^9.

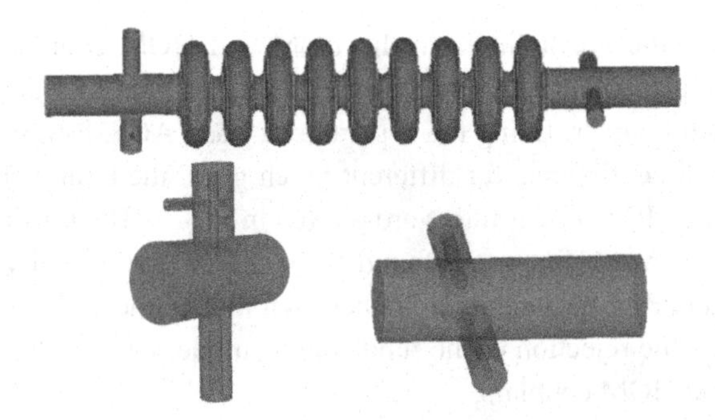

Fig. 6. Model of the ILC crab cavity with modified LOM and HOM couplers.

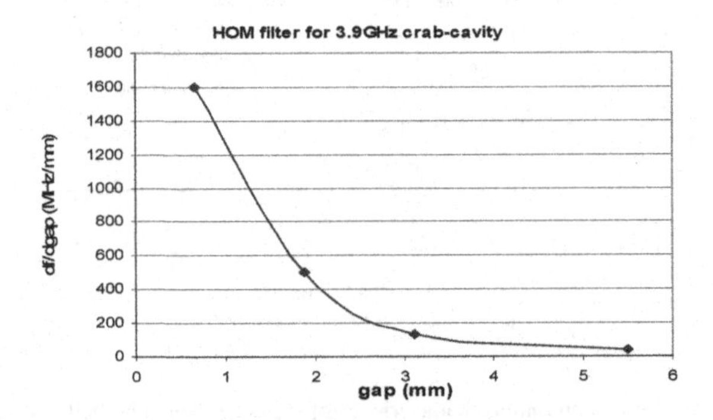

Fig. 7. HOM coupler notch filter sensitivity vs. gap.

The re-design for the HOM coupler is focused on improving the notch filter sensitivity with respect to the gap adjustment. The goal is to reduce the sensitivity from 1.6 MHz/μm to the same level of the TTF cavity, which is 0.1 MHz/μm. The notch frequency of such a coupler is determined by the inductance of the central conductor and the capacitance of the notch gap. The capacitance changes more drastically when the notch gap is small, and therefore it is preferable to have a large gap to reduce the sensitivity. With an increased notch gap, the length of the central conductor needs to be adjusted correspondingly to maintain the notch filter frequency at 3.9 GHz. The filter sensitivity as a function of the gap width is shown in Fig. 7. At a gap width of 3.1 mm, the sensitivity is about 0.1 MHz/μm, which is acceptable.

The new HOM coupler uses a two-stub antenna instead of the coupling loop as shown in Fig. 6. This modification removes the narrow gap between the loop and the outer cylinder and reduces the possibility of multipacting. All the dipole HOMs are well polarized either in the horizontal or vertical plane for the crab cavity. Thus, the HOM coupler can be placed opposite to the input coupler in the horizontal plane to couple out one polarization, and the other polarization can be damped effectively by the LOM and SOM couplers at the other end.

The present SOM coupler is a coax-type coupler similar to the input coupler. Due to the x-y mode mixing, the SOM's Qext is about 5×10^6 for a reasonable central conductor intrusion, and higher than the required 2.6×10^4 from preliminary beam studies [5]. In order to provide sufficient damping for the SOM, the cell indentation is increased (from 1.5 mm to 1.9 mm) to avoid x-y coupling between the SOM and the nearby mode by way of enlarging their mode separation.

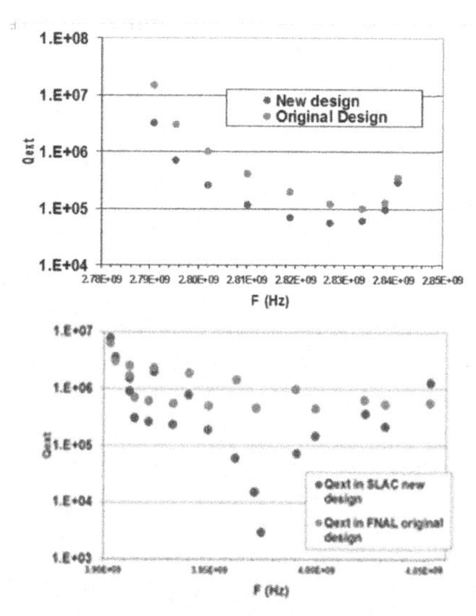

Fig. 8. The damping results of the LOM (up) and HOM (down) from ACE3P-Omega3P for the cavity with new LOM and HOM coupler designs.

Taking into account of the above considerations, the damping of the first monopole and dipole bands calculated using ACE3P-Omega3P are shown in Fig. 8. Significant improvement over the original design can clearly be seen. The SOM's Qext is reduced to an value of 7×10^5, and its mode pattern shown in Fig. 9 shows a pure polarization orthogonal to the operating mode (see top picture in Figure 4). By shaping the tip of the centre conductor of the SOM coupler, the SOM can be damped more effectively. Furthermore, the Qext of the operating mode at the HOM coupler port is found to be 1.4×10^{10}, indicating that it is well rejected by the HOM coupler.

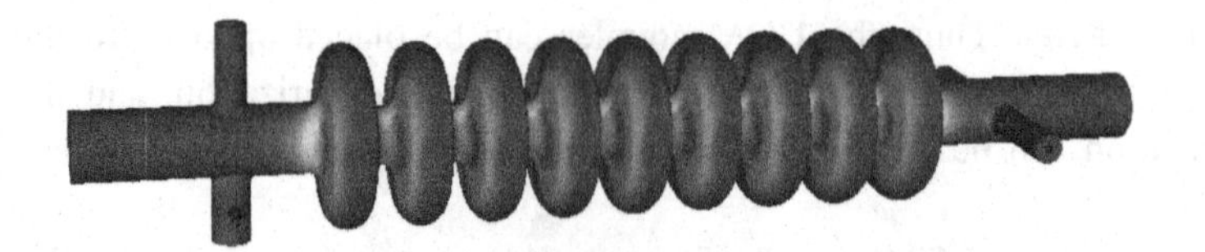

Fig. 9. Field pattern of the SOM in the new crab cavity design.

Modified LOM, HOM and SOM couplers for the ILC crab cavity have been designed and optimized using ACE3P. Prototypes of each of these couplers were manufactured out of copper and measured attached to an aluminum nine cell prototype of the cavity at Cockcroft Institute, Lancaster University, UK. Their external Q factors were measured and found to agree well with ACE3P numerical simulations as shown in Fig. 10.

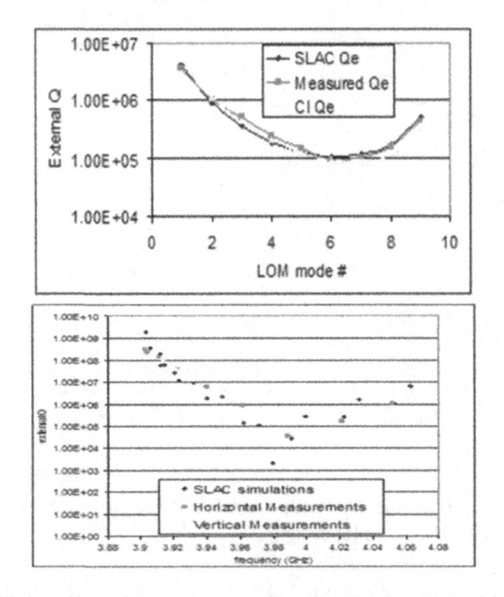

Fig. 10. Comparison of the damping results of the LOM (up) and HOM (down) between measurements and simulations.

4. Multipacting Analysis

Multipacting in the cell and LOM/HOM couplers are simulated using the particle tracking code ACE3P-Track3P. No multipacting activities have been found in the cell and LOM coupler up to a peak transverse field gradient of 5 MV/m. We do find resonant particle trajectories in the HOM pickup region (see Fig. 11). The impact energies of the electrons are between 85 eV and 240 eV, and this energy range is too low to cause multipacting for the copper pickup probe. By rounding the pickup shape with curved surface, the occurrences of resonant trajectories can be reduced.

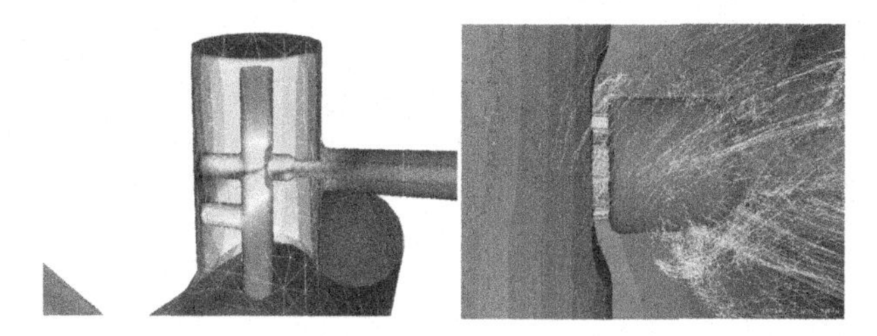

Fig. 11. Resonant particle trajectories in the HOM coupler from ACE3P-Track3P.

5. Summary

The ILC crab cavity is positioned close to the IP and delivered luminosity is very sensitivity to the beam induced HOMs. A set of couplers such as LOM, HOM, and SOM couplers are designed and modified to meet the damping requirements using ACE3P. The modified designs for the LOM and HOM couplers presented in this paper have been manufactured. Their damping results are measured and found to agree well with numerical simulations.

References

1. K. Ko, et al., "Advances in Parallel Electromagnetic Codes for Accelerator Science and development", Proceedings of LINAC2010, Tsukuba, Japan.
2. M. McAshan, R. Wanzenberg, "RF Design of a Transverse Mode Cavity for Kaon Separation", FERMILAB-TM-2144, May 2001.
3. A. Seryi, et al., "Design of the Beam Delivery System for the International Linear Collider", Proceedings of PAC07, Albuquerque, New Mexico, USA.
4. L. Xiao, et al., "HOM/LOM Coupler Study for the ILC Crab Cavity", SLAC-PUB-12409.
5. G. Burt, et al., "Analysis of Damping Requirements for the Dipole Wake-Fields in RF Crab Cavities", IEEE Tran. Nuclear Science. Vol. 54, No. 5. Oct. 2007.
6. G. Burt, et al., "Copper Prototype Measurements of the HOM, LOM, and SOM Couplers for the ILC Crab Cavity", Proceedings of EPAC08, Genoa, Italy.

R&D of C-Band Accelerating Structure at SINAP

Qiang Gu, Wencheng Fang and Zhentang Zhao
FEL Division, Shanghai Institute of Applied Physics, CAS
Shanghai, 201800, P. R. China

Dechun Tong
DEP, Tsinghua University
Beijing, 100018, P. R. China

R&D of a C-band (5712 MHz) high gradient traveling-wave accelerating structure has been in progress at Shanghai Institute of Applied Physics (SINAP). The prototype and new C-band accelerating structures have already been fabricated. In this paper, some details of R&D are introduced.

1. Introduction

Shanghai Soft X-ray FEL facility (SXFEL) has been approved and will start construction soon in 2014 [1]. This facility will be located close to the Shanghai Synchrotron Radiation Facility which is a 3rd generation light source in China [2]. It requires a compact linac with a high gradient accelerating structure and high beam quality. The C-band (5712 MHz) accelerating structure designed to operate at 40 MV/m, is a good choice for a compact linac [3].

In the past several years, several C-band accelerating structure are developed including the prototype and newly optimization structures, and the new structure designed for SXFEL is underway. The prototype structure is based on constant impedance structure, and is used to learn the RF design, fine fabrication, tuning and high power test [4, 5]. Based on the experience of prototype structure, new C-band accelerating structure is optimized with better beam quality, higher power efficiency [6]. According to optimization parameters, one short model has been tuned by cold test and ready for high power as soon as possible. So far almost all technologies of C-band accelerating structure has been developed and ready for final product, and the real product is underway and will be used in 2015 as one of the key components of SXFEL.

In this paper, the process of C-band accelerating structure R&D is introduced, including prototype and new optimization of C-band accelerating structure, and the some details of cold test, high power test results are also presented, and finally the real product design for SXFEL is introduced briefly.

2. Prototype of C-Band Accelerating Structure

The prototype of C-band accelerating structure comprises of regular cells, couplers, high power waveguides and stainless steel flange, which is shown in Fig. 1. The regular cells are conventional disk-loaded type, and the coupler is a two ports electrically-coupled coupler, which can reduce the RF breakdown effectively [7]. The high power waveguides, which are the function of power divider and transmission, are connected by ADESY-type flange. ADESY-type flange performs excellently on microwave transmission compared with the conventional flange [8], and type of flanges for beam tube connection is the CF35.

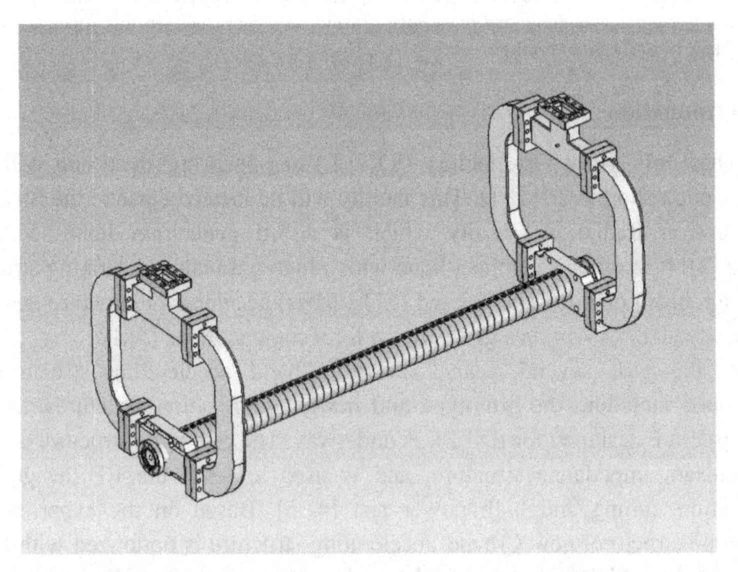

Fig. 1. The schematic drawing of prototype accelerating structure.

Based on the design in Fig. 1, the prototype C-band accelerating structure is fabricated and then cold tested. The tuning bench is shown in Fig. 2. Based on the bead-pull measurement [9], C-band accelerating structure is finely tuned. The tuning results are shown in Fig. 3 and Fig. 4. The field attenuated along the z-axis, and most of the phase shift per cell is below one degree. Since the klystron output power is limited by solenoid mismatch, this C-band accelerating structure has been tested only at 10 MW power, which corresponds to about 20 MV/m.

Fig. 2. The vertical setup for C-band accelerating structure RF cold test.

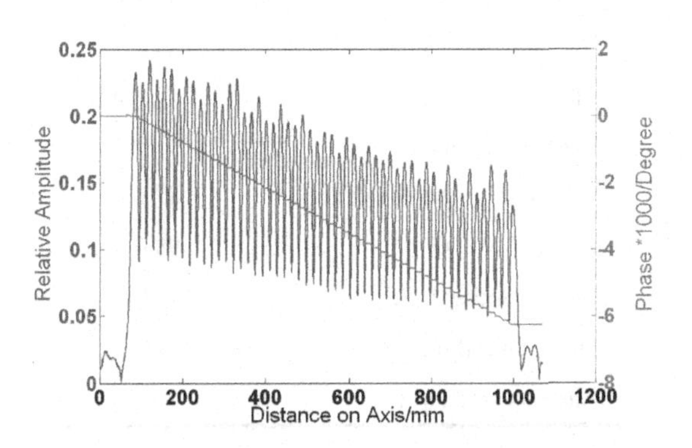

Fig. 3. The field distribution after final tuning.

In all, the designed parameters of the prototype structure are listed in Table 1.

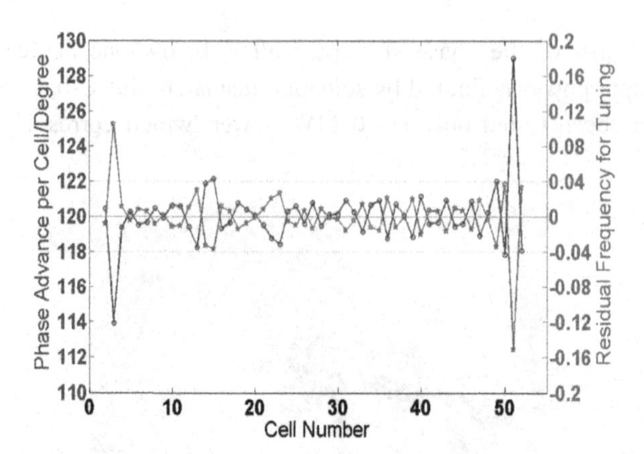

Fig. 4. Phase advance shift per each cell.

Table 1. Parameters of prototype C-band accelerating structure.

Frequency: f(MHz)	5712.000
No. of Cells	53
Phase advance	$2\pi/3$
Total length (mm)	944.73
Length of cell: d (mm)	17.495
Disk thickness: t (mm)	2.500
Aperture: 2a (mm)	10.405
Diameter: 2b (mm)	41.001
Shunt impedance: Rs (MΩ/m)	87.18
Quality factor: Q	9893
Group velocity: Vg/c (%)	1.00
Filling time: tf (ns)	315
Attenuation factor: τ	0.57

3. New C-Band Accelerating Structure

Based on the prototype structure, full optimization has been done on the new structure for the SXFEL facility, including short-term wakefield, peak field and power efficiency, and the parameters of new structure are listed in Table 2.

Table 2. Parameters list of new C-band accelerating structure.

C-band system	SLED-II + CG
Phase advance per cell	$4\pi/5$
Length of cell	20.994 mm
Length of total cells	1.784 m
Average 2a	15 mm
Elliptical tip radius: 2B/2A	9 mm/5 mm
Width of iris: t	5 mm
Peak electric field: E_{peak}/E_0	2.6
Shunt impedance: Rs	Ave: 62 Mohm/m
Q factor	Ave: 10470
Group velocity: vg/c	Ave: 1.7 %
Filling time	340 ns
Attenuation factor: τ	0.585

Before fabrication of the full-size of the structure, one experimental model is accomplished, which is shown in Fig. 5. There are also many improvements of structure, as well as physical parameters, including compact coupler, inner water cooling channel.

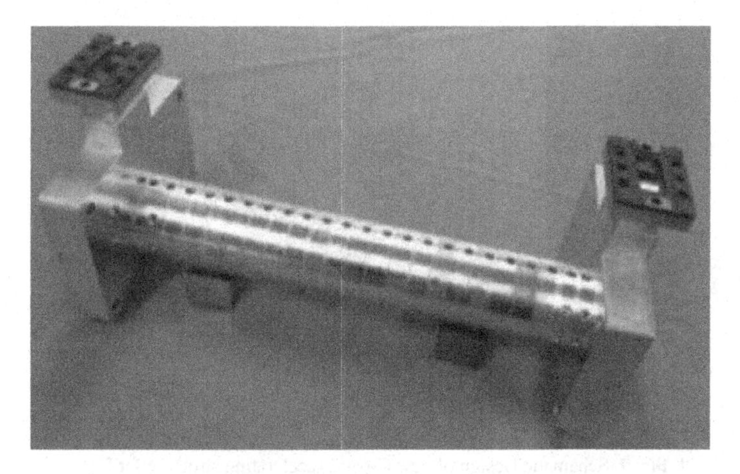

Fig. 5. Experimental model of new C-band accelerating structure.

This constant gradient structure has been tuned carefully, and is ready for high power test soon. The results of tuning are illustrated in Fig. 6. After several times of tuning, the field distribution is almost flat as it should be, and the phase advance per cell is 144 degrees.

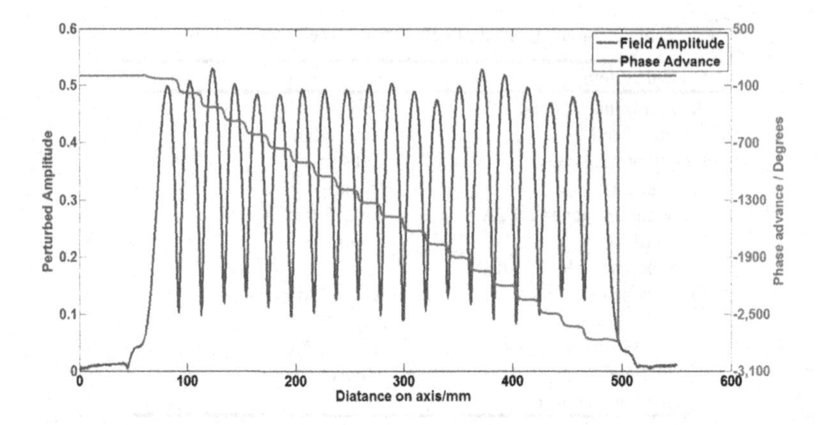

Fig. 6. Field and phase distribution on axis.

Based on the success of experimental model, all key technologies are comprehended, and in the next step, the C band accelerating structure for SXFEL will be fabricated, as shown in Fig. 7.

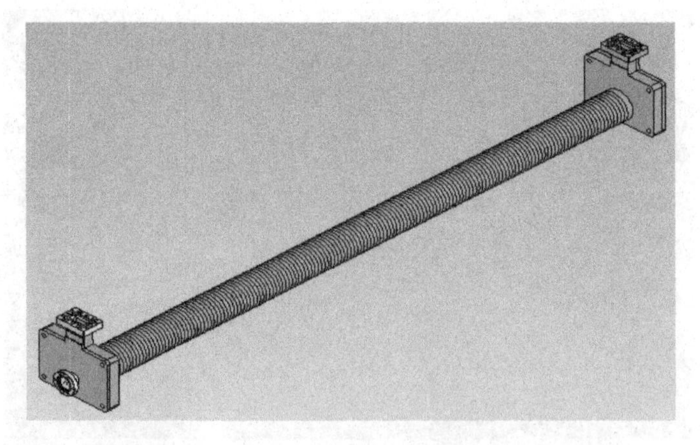

Fig. 7. Schematic Design of new C-band accelerating structure for FEL.

Acknowledgments

It's grateful to Dr. Juwen Wang of SLAC for his helpful suggestions and useful discussion, including high gradient technology, deflecting cavity, pulse compressor and photocathode gun. Juwen is very nice and kind, many people have gained helps from him as well as science and technology, so that this time is good chance to present some work which are related with Juwen, and express our regards and appreciations for him.

References

1. C. Feng *et al.*, Chinese Sci Bull, 2010, Vol. 55 No.3: 221–227.
2. M. Jiang *et al.*, Chinese Sci Bull, 2009, Vol. 54 No.22: 4171–4181.
3. W. Fang *et al.*, Chinese Sci Bull, 2011, Vol. 56 No.1: 18–23 .
4. W. Fang *et al.*, IPAC11, 2011, 133–135.
5. W. Fang *et al.*, Science China Physics, Mechanics & Astronomy 2013 56(11): 2104–2109.
6. W. Fang *et al.*, Chinese Sci bull, 2011, 56(32): 3420–3425.
7. Charles W. Steele, 1966, VOL. MTT-14, NO. 2.
8. T. Khabiboulline *et al.*, 1995, DESY M-95-02.

Juwen and High Gradient Structure Development

From SLAC's SLC to NLC/GLC

R. H. Miller
SLAC National Accelerator Laboratory
Menlo Park, CA 94025, U. S. A.

A Global Effort — The effort to develop high gradient room temperature accelerator structures for the Next Linear Collider or the Global Linear Collider was a global effort involving many people from a number of laboratories around the world including SLAC, FNAL and LLNL in the US, KEK in Japan and CERN in Europe. Critical ideas came from many sources, so my attributions may not always be accurate, but indicate where I first heard the concept. Juwen Wang's role was very important in accepting, modifying where appropriate, and implementing ideas to produce a long series of structures finally leading to a successful structure.

1. Introduction to High Gradient Structures

1.1. *Choice of frequency*

Achievable gradient increases with increasing frequency, but achievable luminosity may decrease. At SLAC; in collaboration with KEK, we chose X band (4 X the SLAC frequency, 2.856 GHZ) for pragmatic reasons. We wanted a harmonic of the SLAC Linac frequency for ease in testing in our existing system. We thought 11.424 GHz might be as high a frequency where we could achieve the desired luminosity and where extrapolations and enhancements of techniques and strategies learned on the existing collider (SLC) would be most useful. Some suggested 17.136 GHz, but the more conservative lower frequency was chosen.

1.2. *Wakefield suppression*

In order for an X band structure to be viable for NLC/GLC the effect of the Dipole Wakefield must be suppressed by about a factor of 100. Heavy damping down to a Q of about 10 would be sufficient. All the heavy damping approaches

Fig. 1. Juwen working on his PHD thesis project in 1983. Here he is studying high gradient performance of an S-Band structure. He also high gradient tested standing wave C-band and X-band structures.

we tried hurt the shunt impedance unacceptably. Juwen pointed out that if the structure was designed to be constant gradient (certainly desirable) the dipole frequencies would vary about 10% from the input end to the output end. This 10% spread in frequency would produce a coherent destructive interference which would for short time reduce the effect of the wakefield by about the same amount as a Q of 10. Over a longer time the interference would lose its coherence so the Q would need to be damped to about 1000. We chose to use a Gaussian density distribution shown in the next figure for the synchronous frequencies for velocity of light electrons of the dipole modes in the cells of the structure. This seemed to give a good reduction of the effect of the dipole mode wakefields on the emittance of the beam. Low beam emittance is necessary for high luminosity in a collider.

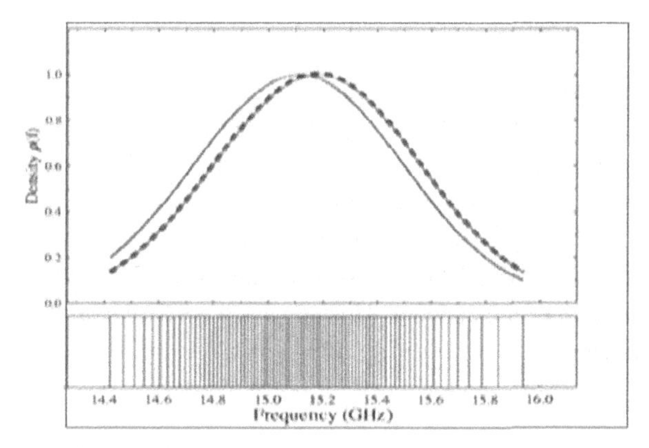

Fig. 2. Gaussian distribution in cell geometries (Solid) and dipole mode frequencies (dotted).

Fig. 3. Juwen looking at first 1.8 m detuned structure.

1.3. *Manifold damping*

Ron Ruth suggested that we might be able to couple the dipole resonances into a manifold running the length of the structure. Robert Gluckstern, visiting from University of Maryland, analyzed the idea and found it promising. Norman Kroll and Roger Jones did a detailed equivalent circuit analyses of the system shown in Fig. 4. They used in their analysis parameter values generated by Zenghai Li using SLAC's EM field solver. Juwen and Zenghai and I designed the implementation of the concept. This approach costs only a few percent in the shunt impedance.

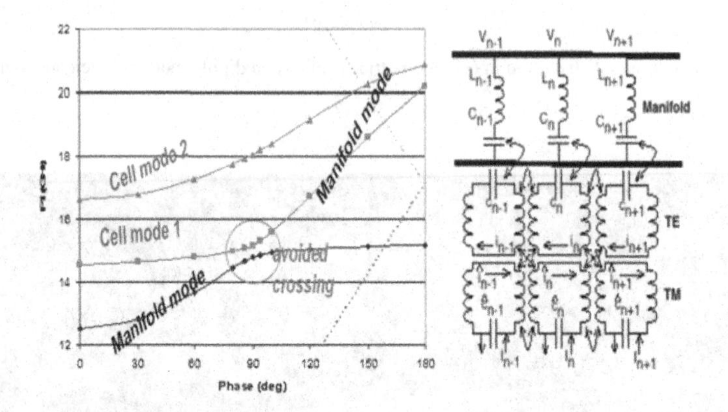

Fig. 4. Dispersion curves for cell modes and manifold modes (R. Jones and N. Kroll).

1.4. *Calculated wakefield*

The equivalent circuit shown in Fig. 4 makes it possible to calculate the wakefields and their effect on the electron beam. Figure 5 (R. Jones) shows how the dipole wakefields due to a single bunch of off-axis electrons decay as a function of distance behind that charge. It shows that electrons about 1 meter behind the driving charge see dipole wakefields reduced by a factor of almost 100.

1.5. *Fabrication*

Early in the SLAC/KEK collaboration on the NLC/GLC linear collider Y. Higashi suggested that we should use diffusion bonding rather than brazing for assembling the accelerator structures, thereby eliminating the detuning effect of alloy fillets. He further suggested precision machining with single crystal

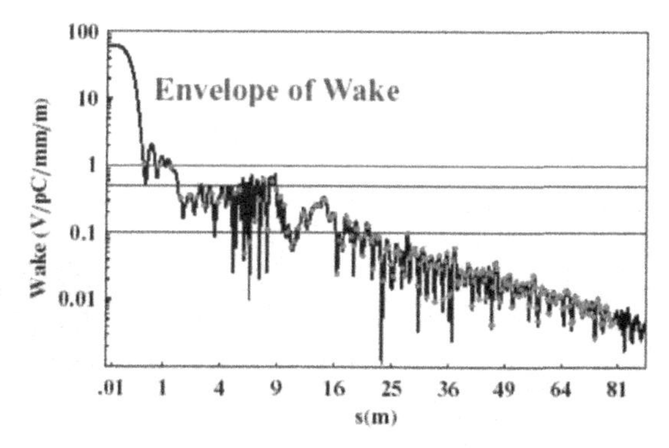

Fig. 5. Calculated wakefield for recently designed HDDS structure. The dots indicate the locations of the bunches, which have a 1.4 ns spacing.

diamond tools in a temperature controlled environment. SLAC followed the first suggestion. Chris Pearson and Juwen worked on an improvement of the technique. Now, SLAC has successfully used diffusion bonding with either single crystal or poly-crystal diamond tools for machining the cells.

Table 1. Typical mechanical tolerances of a structure.

Item	Unit	Specification
Milling positioning	μm	10
Turning diameter	μm	2
Turning depth	μm	2
Tangential discontinuity	degree	5 or 8
Concentricity between milling and turning	μm	10
Concentricity of 2a, 2b and outer diameter	μm	1
Cell-to-cell alignment	μm	5 ~ 10

Fig. 6. One cell of the Damped Detuned Structure. The 4 holes around the cavity are the manifolds.

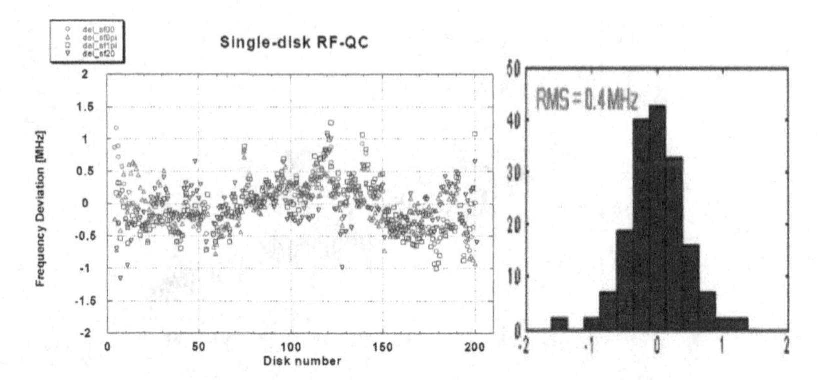

Fig. 7. The measured fundamental and several dipole frequencies for the 201 cells of one of our early Damped Detuned Structures had an RMS deviation of 0.4 MHz from the design values.

2. Achieving High Gradient Operation

I hope I have convinced you that we know how to mitigate the dipole wakefields generated in an X-band structure. Now we must discuss what is involved in achieving high gradient operation in an X-band structure.

We were encouraged by tests in the early 1990's showing that unloaded gradients greater than 100 MV/m are possible in X-band structures. These tests were performed in short standing wave or low group velocity structures in order to get high gradients with the limited X-band power available at the time.

2.1. *Length of accelerator sections*

In the original design an accelerator length of 1.8 m was chosen for pragmatic reasons. This was thought to be a comfortable length to handle and would reduce the number of pieces to be installed. It reduces the number of couplers which are the most complicated part of each structure. This relatively long length for X-Band also simplifies and reduces the cost of the rectangular waveguide drive system. In the SLAC Two Mile Long Linear Accelerator the high power rectangular waveguide system cost about as much as all the linac structures themselves. So reducing the cost of the rectangular waveguide system is a useful goal. However, as we shall see, this economy caused problems.

2.2. *High gradient tests of 1.8m structures*

High Gradient Tests of 1.8 m Structures In 2000–2001, it was discovered that the net rf phase advance through the prototype 1.8 m detuned structures (DS) and damped detuned structures (DDS) had increased by roughly 20 degrees per

1000 hours of operation at gradients as low as 50 MV/m [2]. This was surprising since earlier tests had shown that gradients of more than 80 MV/m could be readily achieved in standing wave and shorter, lower vg traveling wave structures [3]. Table 2 summaries processing of four 1.8 m structures operated at different gradients for different times and the change in the phase advance through the structures caused by arc damage.

Table 2. Structure processing summary.

Structure	Hours Operated	Max Gradient (MV/m)	Phase Change (deg)
DS1	550	54	7
DDS2	550	54	8
DS2	1000	50	20
DDS1	2700	73	60

*This is clearly a disaster!

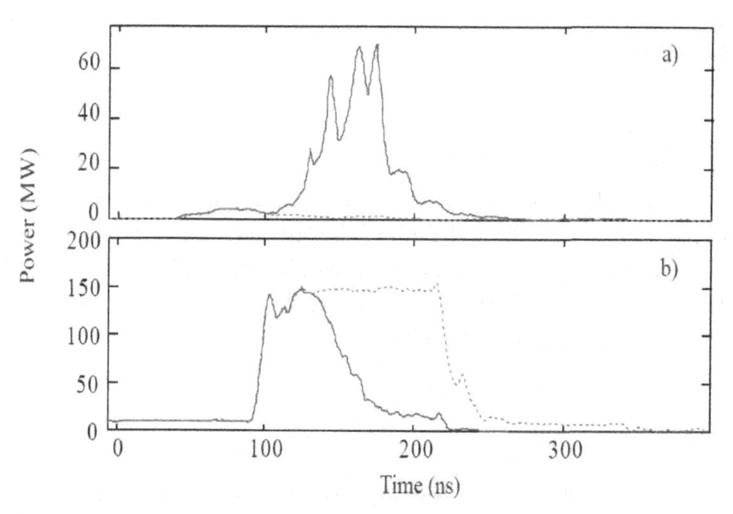

Fig. 8. Example of (a) reflected (b) transmitted power for a breakdown pulse (solid) and normal pulse (dashed).

2.3. *Structure damage increases with higher group velocity*

The previous subsections clearly indicate that most of the breakdowns and most of the resulting damage occur in the higher group velocity part of the structure. Slide 22 shows that the phase error increases with the square or perhaps the cube of the distance up stream of cell 150 where the error vanishes. This suggests that the amount of damage increases between linearly or as the square of distance upstream of cell 150 or roughly linearly or as the square of the increase in group velocity above a threshold group velocity.

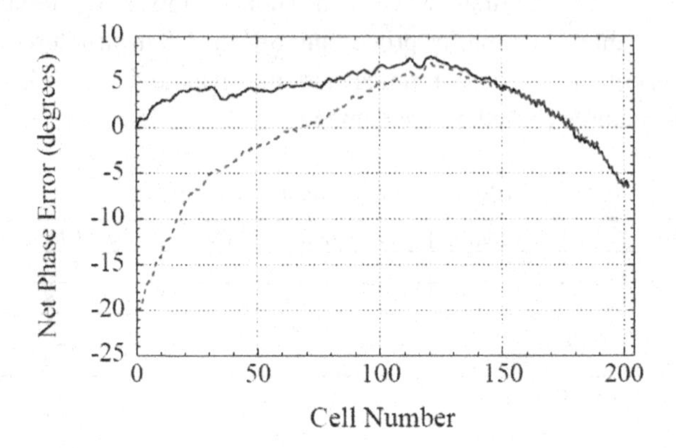

Fig. 9. Bead-pull measurement of the DS2 (third structure in Table 2) phase profile before (solid) and after (dotted) 1000 hours of high power operation.

2.4. *Short low group velocity structures*

The solution was indeed to shorten the structures until the required input power, which increases linearly with the length of the structure and with the square of the desired gradient, was small enough to avoid damaging the structure. The answer for gradients of 65 to 70 MeV/m at 11.424 GHz and our present technology is a structure about 60 cm long.

2.5. *H60VG4SL17*

The length of the structure was cut by a factor of 3 so the group velocity was lowered by a factor of 3 to .04 c and the average aperture was set at $a/\lambda = 0.17$ as assumed in the wakefield calculation. This did not permit the required range of group velocities for a $2\pi/3$ structure, so the group velocities were further lowered by increasing by increasing the phase advance to 5JI/6. The initial H stands for High-Phase-Advance. Roger Jones found that a density distribution proportional to root (sech f) was superior to the Gaussian distribution we had used on the 1.8 meter structure.

Table 3. Basic structures parameters.

Structure name	HDDS (H60VG4SL17)
Structure length	62 cm (including couplers)
Number of acceleration cells	53 cells + 2 matching cells
Average cell iris radius	$\langle a/\lambda \rangle = 0.17$
Phase advance / cell	$5\pi/6$
Group velocity	4.0 ~ 0.9 % speed of light
Attenuation parameter τ	0.64
Filling time	118 ns
Q value	7000 ~ 6500
Shunt impedance	51 ~ 68 MV/m
Coupler	Wave Guide type
1st band dipole mode distribution	Sech distribution with $\Delta f \sim 11\ \%\ (4\sigma)$
Es / Ea	2.22 ~ 2.05
Required input power	59 MW
Gradient without beam $\langle E_0 \rangle$	65 MV/m
Beam loaded gradient $\langle E_L \rangle$	52 MV/m

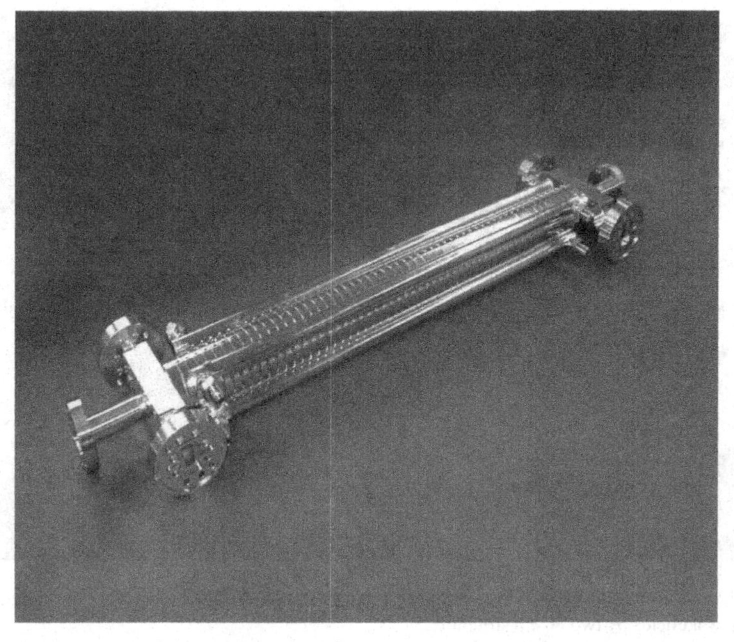

Fig. 10. Nearly final Design Damped Detuned Structure H60VG3S18 with Mode Converter Coupler; Note 4 output couplers at each end from 4 manifolds.

2.6. *ASSET*

Juwen designed and oversaw the fabrication of about 50 structures in this development of room temperature structures for future linear colliders.

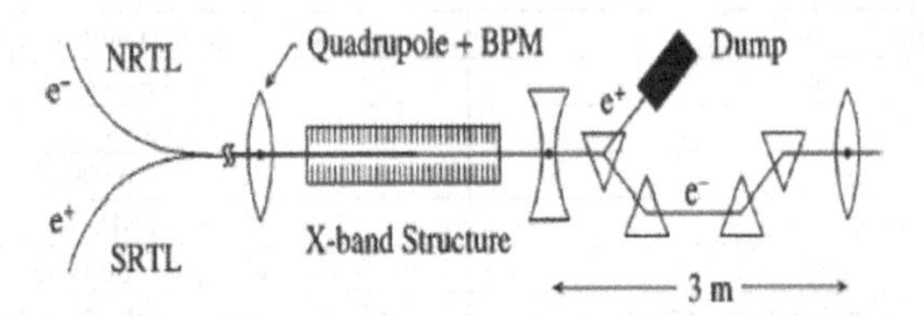

Fig. 11. ASSET is a system designed by C. Adolphsen and installed in the Stanford Linear Collider for measuring the amplitude and time dependence Dipole Wakefields in prospective structures for future linear colliders. The positron bunch serves as a drive beam and the electron bunch as a witness beam.

Fig. 12. Juwen beside ASSET installation of H60vg4SL17A & B for measuring the mitigation of dipole wakefields by two 60 cm structures.

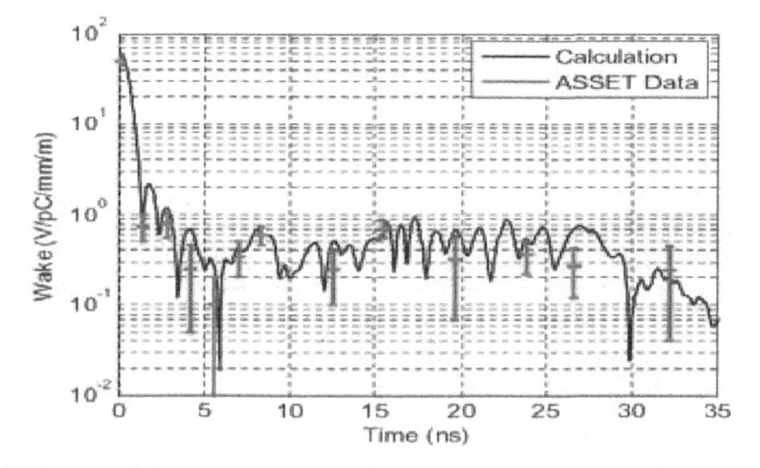

Fig. 13. Comparison of ASSET measurement with error bars (red) and calculated wakefield (black) for a pair of HDDS structures H60VG4SL17A/B. This meets the requirement for Emittance growth in NLC/GLC.

Fig. 14. ILC Test Cavity Team showing 4 of the key players in the High Gradient Structure Task from right: Juwen, Chris Pearson, Chris Adolphsen, and 2nd from left Zenghai Li. Cavity is intended as L-Band Positron Capture section for ILC.

Fig. 15. Juwen, Chris Adolphsen, and Ted Lavine in NLC Test Accelerator.

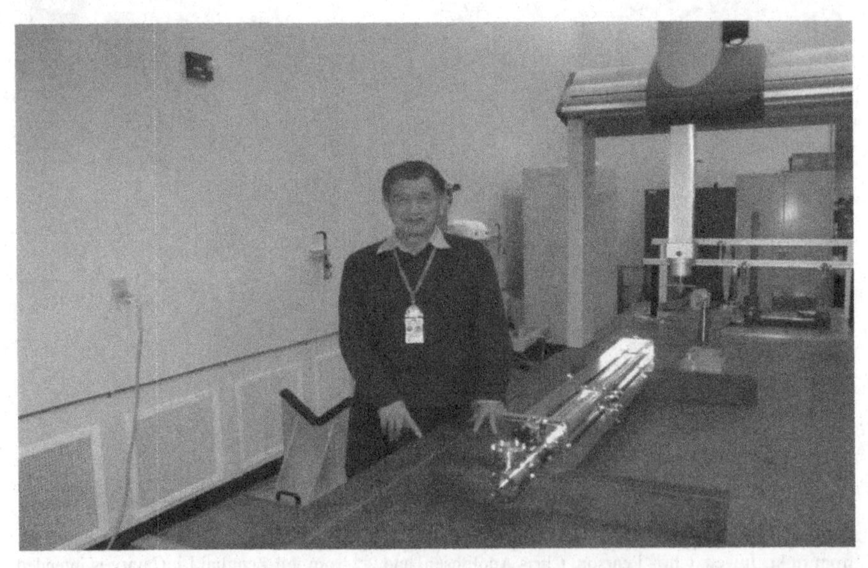

Fig. 16. Juwen with 1 m X-Band RF Deflector. Juwen has made 4 of these and 3 shorter ones for various groups working at SLAC and other labs around the world.

3. Summation

Juwen throughout his career has interacted actively with other accelerator physicists around the world. He has exchanged ideas particularly in the area of high gradient performance and mitigation of dipole wakefields in high frequency accelerator structures. He has listened and assimilated and improved on the concepts and built the structures. Juwen began his work on high gradients 30 years ago in the early 1980's on his PHD thesis project. He has played an absolutely critical role in the development of the NLC/GLC High-phase-advance Damped Detuned Structure (HDDS). For many years he has been my wonderful friend and coworker at SLAC and in the last few years before my retirement, my boss. Through the years we have spent many happy hours discussing the design, optimization, tuning, precision measurement and applications of electron linear accelerators. For a number of years we worked together as consultants at Accuray. We worked both on their medical accelerator and on state of the art accelerators for anti-terrorism scanning of cargo containers, trucks and trains. The anti-terrorism linacs were uniquely capable of switching between multiple energies on a pulse to pulse basis with remarkable energy stability and hence stable X-ray intensity at each energy.

Author Index